Dynamical Scale Transform in Tropical Geometry

Dynamical
Scale Transform in
Tropical Geometry

Tsuyoshi Kato

Department of Mathematics, Kyoto University

 **World Scientific**

NEW JERSEY · LONDON · SINGAPORE · BEIJING · SHANGHAI · HONG KONG · TAIPEI · CHENNAI · TOKYO

Published by

World Scientific Publishing Co. Pte. Ltd.

5 Toh Tuck Link, Singapore 596224

USA office: 27 Warren Street, Suite 401-402, Hackensack, NJ 07601

UK office: 57 Shelton Street, Covent Garden, London WC2H 9HE

Library of Congress Cataloging-in-Publication Data
Names: Kato, Tsuyoshi, 1961–
Title: Dynamical scale transform in tropical geometry / by Tsuyoshi Kato
    (Kyoto University, Japan).
Description: New Jersey : World Scientific, 2016. |
    Includes bibliographical references and index.
Identifiers: LCCN 2016030118 | ISBN 9789814635363 (hardcover : alk. paper)
Subjects: LCSH: Tropical geometry. | Geometry, Algebraic.
Classification: LCC QA582 .K38 2016 | DDC 516.3/5--dc23
LC record available at https://lccn.loc.gov/2016030118

British Library Cataloguing-in-Publication Data
A catalogue record for this book is available from the British Library.

Printed in Singapore

Preface

Tropical geometry is a kind of scale transform which connects a class of rational functions with relative $(\max, +)$-functions. Rational functions appear in almost all mathematical fields, while $(\max, +)$-functions appear in computer science. Even though these functions belong to very different classes, a common feature of functions is that one can obtain orbits by iterating any function. Iterative dynamics is a classical subject and one of the main themes in dynamical systems. Tropical geometry provides us with a new analytic tool to study the structure of such orbits. The basic idea is to extract a dynamical framework from the complicated behavior of the orbits.

Tropical geometry can be widely applied, since both rational and $(\max, +)$-functions are quite broad notions, and they appear in many branches of mathematics. In this book we describe its applications concretely. I have chosen three subjects from projective geometry, infinite groups and mathematical physics. These fields have been individually developed extensively. Tropical geometry presents a new scope to these subjects and builds bridges across materials that are quite different from each other.

Iterative dynamics by rational functions is closely related to analysis of partial differential equations, particularly from the viewpoint of numerical analysis. We also describe the development of a project to construct class analysis of systems of partial differential equations using tropical geometry. Our approach to classification of partial differential equations focuses on the 'rough' global behavior of solutions.

Because the topics treated here are spread across various fields, I have also included explanations of basic things in each subject so that the book is self-contained and accessible to non experts. I have tried to cover the fundamental and core themes of each subject.

This research was partly supported by the Aihara Project, the FIRST program from JSPS, initiated by CSTP, and by the Sakura project from JSPS.

Tsuyoshi Kato

Contents

Mealy type dynamics   199

Chapter 1

Introduction

The theme of this book is to describe geometry and analysis of dynamical systems from the viewpoint of *scale transform*. Dynamical systems can create quite complicated structures, and our aim is to understand them by extracting some simple rules from their framework.

(max, +)-functions have two different aspects, as automata in computer science and as Lipschitz functions in global analysis. The former viewpoint produced several concrete mathematical subjects, while the latter allows us to perform some uniform estimates.

Global analysis given by rational functions plays one of the central subjects in dynamical systems. Let $f_1, f_2, \dots$ be a family of rational functions, and consider their orbits:

$$(f_1(z_1, z_2, \dots), f_2(z_1, z_2, \dots), \dots, f_k(z_1, z_2, \dots), \dots).$$

We will just call them *rational orbits*. In general such orbits behave in quite complicated manners, and direct analysis of them often causes difficulty in understanding their structure.

In some of the rational dynamics, the structure of their orbits contains a 'hidden' framework in a different hierarchy of dynamics. It can be seen by eliminating 'fluctuation' in the orbits, which allows us to develop a systematic study of the classes of dynamics. Let us describe our strategy by the following processes:

(A): Extract simple rules from complex systems.

(B): Compare two complex systems which are reduced to the same rules.

(C): Induce characteristic properties of the simple rules and analyze how such properties are reflected in the original complex systems.

(D): Conversely induce some characteristic properties of the complex systems and analyze how such properties are reflected in the simple rules.

(E): Compare different mathematical objects which arise from very different sources but share some characteristics in their simple rules, and discover structural similarity between them.

It will require using a very strong scaling limit to perform (**A**) above. Actually scale transform can change mathematical structures. Tropical geometry is a kind of dynamical scale transform. The domains change from the positive real line to the real line, and the arithmetics change as:

$$(\times, +) \rightarrow (+, \max)$$

between these semi-rings. The left hand side is the standard one over the real line, and the arithmetic on the right hand side appears in computer science. From the dynamical viewpoint, the scale transform makes all rational orbits degenerate to 0, and the rule or the constraint which governs the dynamics is replaced drastically to a very different one by a change in their arithmetics.

Let $\varphi(y, x)$ be a function of two variables over the real line, and consider the orbits inductively defined by:

$$x_n = \varphi(x_{n-2}, x_{n-1})$$

with initial values $(x_0, x_1) \in \mathbb{R}^2$. This involves iteratively applying φ many times, since $x_2 = \varphi(x_0, x_1)$, $x_3 = \varphi(x_1, x_2) = \varphi(x_1, \varphi(x_0, x_1))$, $x_4 = \varphi(x_2, x_3) = \varphi(\varphi(x_0, x_1), \varphi(x_1, \varphi(x_0, x_1)))$, and so on.

Let us consider examples. Let $\varphi(y, x) = \max(0, y) - x$ be a $(\max, +)$-function. It is not so difficult to check that this dynamics is *recursive*, in the sense that the orbits are periodic with periods 5:

$$x_{n+5} = x_n \qquad \text{for all } n \geq 0$$

with respect to any initial value. The rule in tropical geometry associates the rational function:

$$f(z, w) = \frac{1 + w}{z}.$$

Let us consider the iteration dynamics given by:

$$z_n = f(z_{n-2}, z_{n-1}).$$

It is straightforward to calculate the orbits:

$$z_2 = \frac{1 + z_1}{z_0}, \quad z_3 = \frac{1 + z_0 + z_1}{z_0 z_1}, \quad z_4 = \frac{1 + z_0}{z_1}, \quad z_5 = z_0.$$

So this is also recursive of period 5.

One may imagine that recursiveness could be generally preserved under the scale transform. In fact it is not the case. Let us choose the function:

$$\psi(y, x) = \max(-y, y) - x$$

and consider the iteration dynamics $x_n = \psi(x_{n-2}, x_{n-1})$. It is recursive of period 9. The associated rational dynamics is given by:

$$g(w, z) = \left(w + \frac{1}{w}\right) \frac{1}{z}.$$

It turns out that it does not give recursive dynamics, which can be seen by computing iterations of 9 times with specific initial values.

We will understand such phenomena as *quasi-recursivity* so that its framework consists of recursive dynamics but some fluctuation appears in the rational dynamics, which prevents the rigorous recursivity and is eliminated under the scale transform. It turns out that quasi-recursive dynamics by rational functions exactly correspond to the recursive dynamics by $(\max, +)$-dynamics.

Let us list some of concrete processes we describe in this book. We apply tropical geometry techniques to the theory of pentagram maps which arose from classical projective geometry, the theory of automata groups which consist of a class of infinite groups and KdV equations in integrable systems.

(A): Reduce real rational dynamics to $(\max, +)$-dynamics.

(B): (1) Uniform estimates on the rates of rational orbits.

(2) Rough analytic relation on the set of PDE.

(C): (1) Quasi-recursive rational dynamics is equivalent to the recursive $(\max, +)$-dynamics.

(2) Existence of infinite quasi-recursive rational dynamics from Burnside automaton in group theory.

(3) Perturbation and stationary points; contraction corresponds to boundedness.

(4) Uniform bounds on diameter rates of the invariant tori in the integrable system of the pentagram map.

(D): Duality on automata induced from projective duality over rational functions.

(E): Spectral similarity between KdV and the lamplighter group.

Let us list the subjects we treat in this book along the scale transform, where ∗ are described in a general process which leads to an induction of a class of PDE called hyperbolic Mealy systems:

PDE	Rational dynamics	Automaton
KdV	discrete KdV	BBS_k
∗	∗	lamplighter
∗	rational quasi-Burnside	Burnside
Boussinesq	pentagram map	pentagram

Let us consider the case of group theory. There are important notions which are characteristics in the field, such as amenability, growth function or Dehn function which all measure 'sizes' of some specific qualities of groups. If rational dynamics corresponds to an automata group in tropical geometry, then one wonders how such quantities reflect analytic properties of the rational dynamics or global solutions to the hyperbolic Mealy systems.

As another instance, let us consider integrable systems. As is the case for KdV, scale transform in tropical geometry is fully described from the viewpoint of conserved quantities in integrable systems. However from the classification viewpoint of rational dynamics, equivalent dynamics to these integrable systems are often non integrable, and from a rough analytic viewpoint these non integrable systems still share some kinds of structural similarity with the integrable ones.

One can ask parallel things as above in different situations, which would contain interesting answers.

PART 1
Iterative dynamics

Chapter 2

Tropical transform

2.1 Deformation of semi-ring structure

Tropical mathematics can be regarded as a dequantization of traditional mathematics, whose deformation is known as Maslov dequantization . It transforms standard arithmetics by $(+, \times)$ into one by $(\max, +)$ as the scaling parameter tends to infinity:

$$(+, \times) \to (\max, +).$$

Maslov dequantization leads to an interplay between mathematics over the max-plus algebra (or tropical algebra) and over traditional mathematics.

All these types of algebra admit the structure of *semi-rings*. A semi-ring R is a set equipped with two arithmetics $(*_1, *_2)$ (which correspond to sum and multiplication) such that they satisfy the following properties:

(1) $(R, *_1)$ consists of a commutative monoid with the unit 0,

(2) $(R, *_2)$ consists of a monoid with the unit 1,

(3) $*_2$ is distributive with respect to $*_1$, and

(4) $0 *_2 a = a *_2 0 = 0$.

Let us introduce the structure of a family of semi-rings, which is called the dequantization of the real line $\mathbb{R}$.

Definition 2.1. For $t > 1$, the family of pairs of arithmetics over $\mathbb{R}$, denoted by $(\oplus_t, \otimes_t)$ is given by:

$$x \oplus_t y = \log_t(t^x + t^y), \quad x \otimes_t y = x + y.$$

They are also called the sum and the multiplication.

Lemma 2.1. $(\oplus_t, \otimes_t)$ *is associative for each* $t > 1$. *In particular they consist of a parametrized semi-ring structure over* $\mathbb{R} \cup \{-\infty\}$.

Proof. This can be seen by straightforward calculations. For example:

$$(x \oplus_t y) \otimes_t z = \log_t(t^x + t^y) + z = \log_t[(t^x + t^y)t^z]$$
$$= \log_t(t^{x+z} + t^{y+z}) = (x \otimes_t z) \oplus_t (y \otimes_t z).$$

$\square$

Notice that the above arithmetics $(\oplus_t, \otimes_t)$ is closed under the restriction on $\mathbb{R}$. Let us denote the family by:

$$R_t = \{ \mathbb{R}, \ (\oplus_t, \otimes_t) \ \}.$$

R_t-polynomials over $\mathbb{R}^n$ take the form:

$$\varphi_t(x_0, \ldots, x_{n-1})$$
$$= [\alpha_1 \otimes_t (x_0)^{\otimes_t j_1^0} \otimes_t \cdots \otimes_t (x_{n-1})^{\otimes_t j_1^{n-1}}] \oplus_t \cdots$$
$$\cdots \oplus_t [\alpha_m \otimes_t (x_0)^{\otimes_t j_m^0} \otimes_t \cdots \otimes_t (x_{n-1})^{\otimes_t j_m^{n-1}}]$$

with $\bar{x} = (x_0, \ldots, x_{m-1}) \in \mathbb{R}^n$ and $\bar{j}^l = (j_l^0, \ldots, j_l^{n-1}) \in \mathbb{Z}_{\geq 0}^n$.
Let us rewrite them as:

$$\varphi_t(\bar{x}) = (\alpha_1 + \bar{j}_1 \bar{x}) \oplus_t \cdots \oplus_t (\alpha_m + \bar{j}_m \bar{x})$$

where $\bar{j}\bar{x} = \Sigma_{i=0}^{n-1} j_i x_i$ and $+$ are the usual inner product and the sum respectively.

Relative R_t-polynomials or R_t-rational functions take the form:

$$\varphi_t(\bar{x}) = (\alpha_1 + \bar{a}_1 \bar{x}) \oplus_t \cdots \oplus_t (\alpha_m + \bar{a}_m \bar{x})$$
$$- (\beta_1 + \bar{b}_1 \bar{x}) \oplus_t \cdots \oplus_t (\beta_l + \bar{b}_l \bar{x}) \qquad (*_A)$$

where $\bar{x} = (x_1, \ldots, x_n) \in \mathbb{R}^n$, $\bar{a}_l = (a_l^1, \ldots, a_l^n), \bar{b} \in \mathbb{Z}^n$, $\alpha_i, \beta_i \in \mathbb{R}$.

Later we will understand why we use the term *rational*.

2.2 Tropical algebra

Let us see what happen as $t \to \infty$.

Lemma 2.2. *For any* $(x_0, \ldots, x_m) \in \mathbb{R}^{m+1}$, *the equalities hold:*

$$\lim_{t \to \infty} x_0 \oplus_t x_1 \oplus_t \cdots \oplus_t x_m = \max(x_0, \ldots, x_m).$$

Proof. It is enough to see $x \oplus_\infty y = \max(x, y)$. Suppose $x \geq y$ holds. Then:

$$x \oplus_t y = \log_t(t^x + t^y) = \log_t[t^x(1 + t^{x-y})]$$
$$= x + \log_t(1 + t^{x-y}) \leq x + \log_t 2$$

where the last estimate holds since $x - y \leq 0$. Since $\log_t 2 \to 0$ as $t \to \infty$, and so:

$$x + \log_t 2 \to x = \max(x, y).$$

$\square$

One may guess that the arithmetic $(\max, +)$ also consists of a semi-ring. Actually it is the case.

Definition 2.2. The semi-ring by $\{ \mathbb{R}, (\max, +) \}$ is called tropical algebra.

Polynomials with the coefficient $(\max, +)$ over $\mathbb{R}^n$ take the form:

$$\max(\alpha_1 + \bar{a}_1 \bar{x}, \ldots, \alpha_m + \bar{a}_m \bar{x})$$

which are called $(\max, +)$-functions, where $\bar{x} = (x_1, \ldots, x_n) \in \mathbb{R}^n$, $\bar{a}_l = (a_l^1, \ldots, a_l^n), \bar{b} \in \mathbb{Z}_{\geq 0}^n$, $\alpha_i, \beta_i \in \mathbb{R}$ and $\bar{a}_l \bar{x} = \Sigma_{i=1}^n a_l^i x_i$ are the inner products. They are piecewise-linearly concave functions.

A relative $(\max, +)$-function φ is a piecewise linear function of the form:

$$\varphi(\bar{x}) = \max(\alpha_1 + \bar{a}_1 \bar{x}, \ldots, \alpha_m + \bar{a}_m \bar{x})$$
$$- \max(\beta_1 + \bar{b}_1 \bar{x}, \ldots, \beta_l + \bar{b}_l \bar{x}) \qquad (*_B)$$

where $\bar{a}_s, \bar{b}_t \in \mathbb{Z}^n$. We say that the multiple integer $M \equiv ml$ is the *number of the components* of φ.

Remark 2.1. (1) Later we will use more precise estimates, but it follows from lemma 2.2 that at least these polynomials satisfy 'convergence' in some sense:

$$\lim_{t \to \infty} \varphi_t = \varphi.$$

(2) Notice that the presentations of the relative $(\max, +)$-functions are determined once coefficients $\{\alpha_k, \bar{a}_k, \beta_k, \bar{b}_k\}$ are given.

(3) One can extend the class of functions so that $\bar{a}$ and $\bar{b}$ are in $\mathbb{R}^n$. Actually it causes no difficulty as long as we restrict ourselves to positive real variables when we consider associated rational functions below. Such extensions will be used later.

2.3 Tropical transform

Corresponding to a relative $(\max, +)$-function, tropical geometry associates the parametrized rational function given by:

$$f_t(\bar{z}) \equiv \frac{k_t(\bar{z})}{h_t(\bar{z})} = \left(\Sigma_{k=1}^m t^{\alpha_k} \bar{z}^{\bar{a}_k} \right)\left(\Sigma_{k=1}^l t^{\beta_k} \bar{z}^{\bar{b}_k} \right)^{-1} \qquad (*_C)$$

where $\bar{z}^{\bar{a}} = z_1^{a^1} \ldots z_n^{a^n}$ and $\bar{z} = (z_1, \ldots, z_n) \in \mathbb{R}^n_{>0}$.

We say that f_t above is an *elementary rational function*. Both terms $h_t(\bar{z}) = \Sigma_{k=1}^l t^{\beta_k} \bar{z}^{\bar{b}_k}$ and $k_t(\bar{z}) = \Sigma_{k=1}^m t^{\alpha_k} \bar{z}^{\bar{a}_k}$ are just called elementary functions. Notice that they take positive values over $(0, \infty)^n$.

The fundamental relation in tropical geometry is given by the following. Let:

$$\mathrm{Log}_t : \mathbb{R}^n_{>0} \to \mathbb{R}^n \quad (x_1, \ldots, x_n) = (\log_t z_1, \ldots, \log_t z_n)$$

and φ_t, f_t be $*_A$ and $*_C$ as above.

Proposition 2.1. *The conjugation by $\log_t$ of φ_t as:*

$$f_t \equiv (\log_t)^{-1} \circ \varphi_t \circ \mathrm{Log}_t : \mathbb{R}^n_{>0} \to (0, \infty)$$

is the parametrized rational function $f_t(\bar{z}) = \frac{\Sigma_{k=1}^m t^{\alpha_k} \bar{z}^{\bar{a}_k}}{\Sigma_{k=1}^l t^{\beta_k} \bar{z}^{\bar{b}_k}}.$

Proof. Notice the following three formulas:

(1) $\bar{a} \, \mathrm{Log}_t(\bar{z}) = \Sigma_{i=1}^n a^i \log_t(z_i) = \Sigma_{i=1}^n \log_t(z_i^{a^i}) = \log_t(\bar{z}^{\bar{a}})$,

(2) $\log_t(a) \oplus_t \log_t(b) = \log_t(t^{\log_t(a)} + t^{\log_t(b)}) = \log_t(a + b)$,

(3) $\log_t(f_t(\bar{z})) = \log_t(\Sigma_{k=1}^m t^{\alpha_k} \bar{z}^{\bar{a}_k}) - \log_t(\Sigma_{k=1}^l t^{\beta_k} \bar{z}^{\bar{b}_k})$.

Then we obtain the equalities by applying the above formulas:

$$\varphi_t(\mathrm{Log}_t(\bar{z})) = (\alpha_1 + \bar{a}_1 \mathrm{Log}_t(\bar{z})) \oplus_t \cdots - (\beta_1 + \bar{b}_1 \mathrm{Log}_t(\bar{z})) \oplus_t \ldots$$
$$= (\log_t(t^{\alpha_1}) + \log_t(\bar{z}^{\bar{a}_1})) \oplus_t \ldots$$
$$- (\log_t(t^{\beta_1}) + \log_t(\bar{z}^{\bar{b}_1})) \oplus_t \ldots$$
$$= \log_t(t^{\alpha_1} \bar{z}^{\bar{a}_1}) \oplus_t \cdots - \log_t(t^{\beta_1} \bar{z}^{\bar{b}_1}) \oplus_t \ldots$$
$$= \log_t(\Sigma_{k=1}^{m} t^{\alpha_k} \bar{z}^{\bar{a}_k}) - \log_t(\Sigma_{k=1}^{l} t^{\beta_k} \bar{z}^{\bar{b}_k}).$$

$\square$

So far we have considered three different classes of functions φ, φ_t and f_t, which admit one to one correspondences among their presentations once coefficients $\{\alpha_k, \bar{a}_k, \beta_k, \bar{b}_k\}$ are given (see remark 2.1(2)):

$$(*_A) \quad \varphi_t(\bar{x}) = (\alpha_1 + \bar{a}_1\bar{x}) \oplus_t \cdots \oplus_t (\alpha_m + \bar{a}_m\bar{x})$$
$$- (\beta_1 + \bar{b}_1\bar{x}) \oplus_t \cdots \oplus_t (\beta_l + \bar{b}_l\bar{x})$$
$$(*_B) \quad \varphi(\bar{x}) = \max(\alpha_1 + \bar{a}_1\bar{x}, \ldots, \alpha_m + \bar{a}_m\bar{x})$$
$$- \max(\beta_1 + \bar{b}_1\bar{x}, \ldots, \beta_l + \bar{b}_l\bar{x})$$
$$(*_C) \quad f_t(\bar{z}) = (\Sigma_{k=1}^{m} t^{\alpha_k} \bar{z}^{\bar{a}_k})(\Sigma_{k=1}^{l} t^{\beta_k} \bar{z}^{\bar{b}_k})^{-1}$$

where $(*_A)$ approaches $(*_B)$ as $t \to \infty$ by lemma 2.2, and $(*_A)$ and $(*_C)$ are mutually conjugate by $\log_t$ by proposition 2.1.

Definition 2.3. For given coefficients $\{\alpha_k, \bar{a}_k, \beta_k, \bar{b}_k\}$, we call the corresponding triplet $\{\varphi_t, \varphi, f_t\}$ above the *tropical triplet*.

They are mutually called tropical correspondences.

Sometimes we may drop t and just denote f for elementary rational functions.

We have given a process of scale transform in tropical geometry, from elementary rational functions to relative $(\max, +)$-functions:

$$f_t \Rightarrow \varphi$$

passing through some intermediate functions φ_t.

As $t \to \infty$, φ_t approaches φ, but the image of Log_t contracts quite strongly so that limit of Log_t degenerates and just the constant.

Let $(\varphi_t, \varphi, f_t)$ and (ψ_t, ψ, g_t) be tropical triplets. Then the following is easy to check:

Lemma 2.3. *(1)* $\max(\varphi, \psi)$ *corresponds to* $f_t + g_t$.
  (2) $\varphi + \psi$ *corresponds to* $f_t g_t$.
  (3) $-\varphi$ *corresponds to* f_t^{-1}.

So $N f_t$ and $L^{-1} f_t$ correspond to $\max(\varphi, \ldots, \varphi)$ and $\varphi - \max(0, \ldots, 0)$ respectively. Both the latter coincide with φ as functions, but their presentations are all different. This viewpoint leads to a notion of tropical equivalence introduced below.

2.4 Uniqueness of the scale transform

One may think that tropical geometry could be a special instance among various scale transforms. However it turns out that tropical geometry satisfies some universality among scale transforms from the arithmetic viewpoint.

Lemma 2.4. *Let* $f : \mathbb{R} \to \mathbb{R}$ *be a continuous map with the property:*
$$f(x + y) = f(x) f(y), \qquad x, y \in \mathbb{R}.$$
Then there is some $t > 0$ *so that one of* $f(x) = t^x$ *or* $f(x) = 0$ *holds.*

Proof. We claim that $f(a) = (f(1))^a$ holds for any rational $a = \frac{m}{n} \in \mathbb{Q}$. Suppose $f(1) = 0$ holds. Then it follows from the equalities $f(1) = f(n\frac{1}{n}) = f(\frac{1}{n})^n$ that $f(\frac{1}{n}) = 0$ also holds. Then $f(a) = f(\frac{1}{n})^m = 0$ also holds.
  If $f(1) \neq 0$ holds, then positivity $t \equiv f(1) > 0$ must hold since $f(1) = f(\frac{1}{2})^2 > 0$. Then the equality holds:
$$f(a) = f\left(\frac{1}{n}\right)^m = f(1)^{\frac{m}{n}} = f(1)^a.$$
Thus $f(a) = t^a$ holds for any $a \in \mathbb{R}$ by continuity. $\qquad\square$

2.5 Intermediate selection

Let $a, b, c \in \mathbb{R}$, and notice:
$$\min(a, b) = -\max(-a, -b).$$
It is immediate to check the equalities:
$$a = \max(a, \min(a, b)) = \max(a, -\max(-a, -b)).$$

The tropical correspondence to the right hand side is given by:

$$z + \frac{zw}{z+w}$$

which is different from z.

According to D.Takahashi, the following number gives the intermediate number:

$$m(a,b,c) \equiv \max(\min(a,b), \min(b,c), \min(c,a))$$

so that it picks up b if $a \le b \le c$ holds.

Its tropical correspondence is given by:

$$\left(\frac{1}{z} + \frac{1}{u}\right)^{-1} + \left(\frac{1}{u} + \frac{1}{w}\right)^{-1} + \left(\frac{1}{w} + \frac{1}{z}\right)^{-1}.$$

In a large scale u should behave 'larger' than z and 'smaller' than w.

So the estimates hold:

$$\min(a,b,c) \le m(a,b,c) \le \max(a,b,c).$$

The analysis of this operation is beyond the scope of this book, and will not treated here.

2.6 Tropical equivalence

Let us start with an example. Consider a $(\max, +)$-function over $\mathbb{R}^2$:

$$\varphi(x,y) = \max(-y, y) - x.$$

Because one of y or $-y$ must be non negative, $\max(-y, y) \ge 0$ always holds. In particular, the equality holds as **functions**:

$$\varphi(x,y) = \psi(x,y) \equiv \max(\varphi(x,y), -x).$$

Such functions φ and ψ are called *equivalent*.

On the other hand φ and ψ have mutually different presentations, and as a result, they have different tropical correspondences:

$$f(z,w) = (w^{-1} + w)z^{-1}, \quad g(z,w) = (w^{-1} + w + 1)z^{-1}.$$

f and g are different even as functions.

So different rational functions can arrive at the same $(\max, +)$-functions passing through the scaling limit.

Definition 2.4. Let f and g be two elementary functions, and let φ and ψ be their tropical correspondences.

f and g are tropically equivalent, if φ and ψ are the same as functions but possibly have different presentations.

2.7 Examples

Let φ and ψ be relative $(\max, +)$-functions, and f and g be their tropical correspondences.

(1) If $\psi = \max(\varphi, \varphi)$, then $g = 2f$. φ and ψ are the same functions, but have different presentations. So f and $2f$ are tropically equivalent. More generally f and nf are tropically equivalent for any $n \in \mathbb{N}$.

If $\psi = \varphi - \max(0, \ldots, 0)$, then $g = \frac{1}{m}f$. Of course φ and ψ are the same functions and so f and $\frac{1}{m}f$ are mutually tropically equivalent for any $m \in \mathbb{N}$. Combining with these, we see f and αf are mutually tropically equivalent for any positive rational number $\alpha \in \mathbb{Q}_{>0}$.

(2) Suppose the inequality $\varphi \geq \psi$ holds pointwisely. If $\varphi' \equiv \max(\varphi, \psi)$, then the tropical correspondence to φ' is $f + g$. f and $h \equiv f + g$ are mutually tropically equivalent, since φ and φ' are the same functions.

For example let us apply this to $\varphi(x) = \max(-x, x)$ and $\varphi'(x) = \frac{1}{2}x$. Then $f(z) = z^{-1} + z$ and $h(z) = z^{-1} + z + \sqrt{z}$ are mutually tropically equivalent.

Let f be an elementary rational function and $[f]$ be its tropically equivalent classes. It is immediate to see that the set of tropically equivalent classes forms a semi-ring:

$$\widetilde{\mathbf{TP}} = \{[f] : f : \text{elementary rational functions}\}$$

by $[f][g] = [fg]$ and $[f] + [g] = [f + g]$. Notice that positive rational numbers $\mathbb{Q}_{>0}$ act on $\widetilde{\mathbf{TP}}$ trivially.

Its Grothendieck ring is given by:

$$\mathbf{TP} = \text{gen } \{[f] : \text{elementary rational functions}\}$$
$$\{[f][g] = [fg], [f] + [g] = [f + g]\}.$$

So far the algebraic structure of $\mathbf{TP}$ has not yet been studied extensively.

Problem 2.1. *Study the semi-ring structure of* $\mathbf{TP}$.

2.8 Projective duality

Let us introduce a new duality on the set of relative $(\max, +)$-functions. It passes through the *projective duality* between algebraic varieties. This induces a duality on some classes of automata which include Mealy type automata which we will treat later.

Let V be an n dimensional vector space and $P(V)$ be its projective space. There is a natural isomorphism $P(V) \cong P^*(V^*)$, where $P^*(W)$ is the set of all hyperplanes in W, and V^* is the dual space to V.

Projective duality generalizes this operation to varieties. Let us quickly describe its construction. Let $X \subset P(V)$ be an algebraic variety. Then one can associate another variety $X^\vee \subset P(V^*)$ as follows. A hyperplane $H \subset P(V)$ is said to be tangent to X, if there exists a smooth point $x \in H \cap X$ and the tangent space of X at x is contained in H. Let $X^* \subset P^*(V)$ be all the set of all tangent hyperplanes, and passing through the above isomorphism, one obtains a set $X^\vee \subset P(V^*)$ which is the desired one. It is called the *projective dual variety*. In the case when $X^\vee \subset P(V^*)$ is a hypersurface, then its defining polynomial Δ_X is called the X-*discriminant*.

Let $\{\varphi_1, \ldots, \varphi_m\}$ be a family of relative $(\max, +)$-functions, and denote the corresponding elementary rational functions by $\{f_t^i\}_{i=1}^m$. Thus one obtains a parametrized family of projective varieties $V(\{f_t^i\})_t$ by taking closure of the associated affine varieties.

Let $V(\{f_t^i\})_t^\vee$ be the corresponding parametrized projective dual varieties. When the defining functions $\{g_t^1, \ldots, g_t^l\}$ are given by elementary rational functions, the corresponding family of relative $(\max, +)$-functions $\{\varphi_1^\vee, \ldots, \varphi_l^\vee\}$ is called the projective dual of $\{\varphi_1, \ldots, \varphi_m\}$. When $m = 1$, then we call $\varphi^\vee$ the discriminant.

Suppose $\{\varphi_1, \ldots, \varphi_m\}$ is restricted to the family of maps between the same finite subset in the Euclidean space and denote the restriction by $\mathbf{A}$. We say that $\{\varphi_1^\vee, \ldots, \varphi_l^\vee\}$ is restricted to a *dual automaton* $\mathbf{A}^\vee$, if it is also restricted to the family of maps between the same finite subset in the Euclidean space.

2.9 Curves in $\mathbb{C}P^2$

In general it is not easy to calculate the defining equations of the projective varieties. Let us compute it for a particular case.

Lemma 2.5. *The discriminant is given as:*

$$[\max\{au, \alpha + av\} = c]^\vee =$$
$$\max\left\{ \frac{a}{a-1}\left(c - \frac{\alpha}{a}\right) + \frac{a}{a-1}v, \ \frac{ac}{a-1} + \frac{a}{a-1}u \right\} = c.$$

Proof. Let $X \subset \mathbb{C}P^2$ be an irreducible curve. Then $X^\vee \subset \mathbb{C}P^2$ is also another irreducible one. On the affine coordinate, if X has a parametrization

$x = x(s)$ and $y = y(s)$, $s \in \mathbb{C}$, then $X^\vee$ has a parametrization given by the following:

$$p(s) = \frac{-y'(s)}{x'(s)y(s) - x(s)y'(s)}, \quad q(s) = \frac{x'(s)}{x'(s)y(s) - x(s)y'(s)}.$$

Using this, let us consider the very simple case above. Let $a \geq 2$, α and c be integers, and consider an automaton given by:

$$\max\{au, \alpha + av\} = c.$$

The associated polynomial and the associated varieties are given by:

$$X = \{(x,y) \in \mathbb{C}^2 \subset \mathbb{C}P^2 : x^a + t^\alpha y^a = t^c\}.$$

Choosing a parametrization as:

$$x = s, \quad y = t^{-\frac{\alpha}{a}}(t^c - s^a)^{\frac{1}{a}}$$

one can immediately obtain the parametrization of $X^\vee$ as:

$$t^{\frac{a}{a-1}(c - \frac{\alpha}{a})}q^{\frac{a}{a-1}} + t^{\frac{ac}{a-1}}p^{\frac{a}{a-1}} = t^c \quad (*)$$

which gives the dual varieties:

$$V(A)_t^\vee = \{(p,q) \in \mathbb{C}^2 \subset \mathbb{C}P^2 : (*)\}.$$

Thus the projectively dual automaton admits the desired presentation.   □

References

Tropical geometry is now a popular subject and the topics in sections 2.1 ~ 2.3 are quite standard (see [LM],[Mik], [Vir] for basic information). Its scaling limit has also been used in integrable systems, which is known as ultradiscrete.

There are many references on ultradiscrete integrable systems. See [TTMS] for example. Tropical equivalence is closely related to the ideas of compactification of metric spaces [Gro1]. But its convergence is much stronger, since it is the space of the defining equations rather than the orbits themselves that are compactified [Kat4]. [GKZ] is the basic textbook on projective duality. See also [Kat6].

Chapter 3

Dynamical hierarchy

3.1 Iterated dynamics

Let $\varphi : \mathbb{R}^n \to \mathbb{R}$ be a function. For any initial value $\bar{x}_0 = (x_0, \ldots, x_{n-1}) \in \mathbb{R}^n$, let us consider its orbit:

$$x_0, x_1, \ldots, x_m, x_{m+1}, \ldots$$

which is inductively defined by:

$$x_{m+1} = \varphi(x_{m-n+1}, \ldots, x_m) \qquad (m \geq n - 1).$$

This dynamics is determined uniquely by the initial value. We call such a property deterministic. There are other important dynamics which are not deterministic, which we will not treat in this book.

In order to trace the orbit, one has to compute the iterated values of φ. In fact:

$$x_n = \varphi(\bar{x}_0),$$
$$x_{n+1} = \varphi(x_1, \ldots, x_n) = \varphi(x_1, \ldots, \varphi(\bar{x}_0)),$$
$$x_{n+2} = \varphi(x_2, \ldots, x_n, x_{n+1}) = \varphi(x_2, \ldots, x_{n-1}, \varphi(\bar{x}_0), \varphi(x_1, \ldots, \varphi(\bar{x}_0))),$$
$$x_{n+3} = \varphi(x_3, \ldots, x_{n-1}, \varphi(\bar{x}_0), \varphi(x_1, \ldots, \varphi(\bar{x}_0)),$$
$$\varphi(x_2, \ldots, x_{n-1}, \varphi(\bar{x}_0), \varphi(x_1, \ldots, \varphi(\bar{x}_0)))),$$

$$\cdots$$

Remark 3.1. In order to determine x_N, one needs to iterate $N - n + 1$ times to apply function φ.

What characteristics of dynamics can you analyze? The simplest one will be a fixed point. You can find it by solving the equation $x_0 = \varphi(\bar{x}_0)$. It is of interest to pursue how a neighborhood of fixed points behave under

17

the same dynamics, or how the points behave under a small perturbation of the defining equations. Later we will treat the latter case and see how the orbits change under tropical scale transform.

The next characteristic itself is quite interesting, which involves periodic orbits.

Definition 3.1. φ is recursive of period $p \geq n$, if the equalities:

$$x_{p+N} = x_N \in \mathbb{R} \qquad (N \geq 0)$$

hold for any initial value $(x_0, \ldots, x_{n-1}) \in \mathbb{R}^n$ on the domain.

Lemma 3.1. *Recursivity is equivalent to a statement that $x_0 = x_p$ holds for any initial value $(x_0, \ldots, x_{n-1}) \in \mathbb{R}^n$.*

Proof. Recursivity implies $x_0 = x_p$ for any initial value $(x_0, \ldots, x_{n-1}) \in \mathbb{R}^n$ by definition.

Conversely if $x_0 = x_p$ hold for any initial values $(x_0, \ldots, x_{n-1}) \in \mathbb{R}^n$, then $x_1 = x_{p+1}$ also holds since such a property is independent of choice of the initial values. One can iterate to see recursivity. $\qquad \square$

When φ is a relative $(\max, +)$-function, then we can consider the same notion of recursiveness for tropical correspondences φ_t and f_t, where their domains are $\mathbb{R}^n$ and $\mathbb{R}^n_{>0} = (0, \infty) \times \cdots \times (0, \infty)$ respectively, and we assume for all $t > 1$ otherwise stated.

Also we assume that the period p is minimal unless otherwise stated.

Example 3.1. (1) Let $\varphi(x, y) = \max(0, y) - x$. It is recursive of period 5. Its tropical correspondence $f(z, w) = (1 + w)z^{-1}$ is also recursive of the same period.

(2) Let $\varphi(x, y) = \max(-y, y) - x$. It is recursive of period 9. Its tropical correspondence $f(z, w) = (w + w^{-1})z^{-1}$ is not recursive.

(3) Let $\varphi(x, y) = 1 + y - x$. It is recursive of period 6. Its tropical correspondence $f_t(z, w) = t\frac{w}{z}$ is also recursive of the same period.

These examples imply that recursivity is not an inheritable trait under tropical transform. Later we will find a correct notion over f_t corresponding to recursive φ.

3.2 Dynamical hierarchy

Let $\{\varphi, \varphi_t, f_t\}$ be a tropical triplet. Recall that they admit one to one correspondence of their presentations which are determined by the coefficients $\{\alpha_k, \bar{a}_k, \beta_k, \bar{b}_k\}$ in 2.3. It would be natural to study the dynamical systems of different scales at the same time:

$$x_{m+1} = \varphi(x_{m-n+1}, \ldots, x_m),$$
$$x'_{m+1} = \varphi_t(x'_{m-n+1}, \ldots, x'_m),$$
$$z_{m+1} = f_t(z_{m-n+1}, \ldots, z_m)$$

with the initial values as:

$$x_i = x'_i = \log_t z_i \quad (0 \le i \le n-1).$$

Notice that the orbits $\{x'_i\}$ and $\{z_i\}$ both depend on the parameter $t > 1$, but we will omit denoting it explicitly.

Lemma 3.2. $x'_m = \log_t z_m$ for all $m \ge 0$.

Proof. It follows from proposition 2.1 that f_t and φ_t are conjugated by $\log_t$. Hence the equality holds:

$$t^{x'_{m+1}} = t^{\varphi_t(x'_{m-n+1}, \ldots, x'_m)}$$
$$= f_t(t^{x'_{m-n+1}}, \ldots, t^{x'_m}).$$

Since the initial value is given by $z_0 = t^{x'_0}, \ldots, z_{n-1} = t^{x'_{n-1}}$, the conclusion follows, since the orbit is uniquely determined by the initial value. $\qquad\square$

Remark 3.2. (1) It follows from this lemma that some important properties of the deterministic dynamics by φ_t and f_t are shared by this conjugacy such as recursivity or existence of periodic orbits.

(2) At a glance x_m and x'_m seem very close from each other for large $t \gg 1$ (see remark 2.1). In fact each x'_m converges to x_m as $t \to \infty$. However such convergence is not uniform with respect to m, in general. Otherwise such phenomena as example 3.1(2) would not happen by the above lemma.

Recall that we have introduced tropical equivalence between elementary rational functions f and g. There is an important notion in dynamical systems as follows. Let us say that f and g are *conjugate*, if there is a one to one onto map $h : (0, \infty) \cong (0, \infty)$ so that the equality:

$$h \circ f(z_0, \ldots, z_{n-1}) = g(h(z_0), \ldots, h(z_{n-1}))$$

holds for all $(z_0, \ldots, z_{n-1}) \in \mathbb{R}^n_{>0}$.

Particularly topological conjugacy plays an important role in the study of dynamical systems, since many dynamical properties are preserved under change by topological conjugacy.

Proposition 3.1. *Tropical equivalence does not imply conjugacy in general.*

Proof. Example 3.1(1) can produce a particular case. Let $f(z, w) = (1 + w)z^{-1}$ be the elementary rational function, which gives the recursive dynamics.

Let $g = 2f$ which is tropically equivalent to f. It can be checked straightforwardly that g does not give recursive dynamics.

Since recursiveness is preserved under conjugacy, this pair of the example gives the conclusion. □

An interesting question arises from the above proposition. Let:

$$f(z, w) = (w^{-1} + w)z^{-1}$$

be the elementary rational function which gives non recursive dynamics in the example 3.1(2).

Question 3.1. *Can one find recursive functions among the tropically equivalent class of f?*

This is not known at present.

3.3　Basic estimates

Let us introduce the metric on $\mathbb{R}^n$ given by:

$$d((x_0, \ldots, x_{n-1}), (y_0, \ldots, y_{n-1})) \equiv \max_{0 \leq i \leq n-1} \{|x_i - y_i|\}.$$

This is equivalent to the Euclidean metric in the sense that there are constants C_n, C'_n which depends only on n such that the estimates hold:

$$C_n^{-1} \sqrt{\Sigma_{i=0}^{n-1} |x_i - y_i|^2} \leq \max_{0 \leq i \leq n-1} \{|x_i - y_i|\} \leq C'_n \sqrt{\Sigma_{i=0}^{n-1} |x_i - y_i|^2}.$$

In this particular case, one can in fact choose $C_n = \sqrt{n}$ and $C'_n = 1$. So it will not change the 'rough' metric structure from the Euclidean metric. Such equivalence does not hold over the infinite dimensional Hilbert spaces which correspond to l^1 and l^2 metrics over $\mathbb{N}$ respectively, and so finite dimensionality is necessary for the equivalence.

The key notion to the global analysis of the orbits is *Lipschitz property*. The Lipschitz constant $c \geq 0$ of a function φ is given by a number which admits the inequality for all $\bar{x}, \bar{y} \in \mathbb{R}^n$:

$$|\varphi(\bar{x}) - \varphi(\bar{y})| \leq c \, d(\bar{x}, \bar{y}).$$

Remark 3.3. (1) In our case the global Lipschitz property is more important than the local one, so that differentiable functions are not necessarily Lipschitz. For example $f : \mathbb{R} \to \mathbb{R}$ by $f(x) = x^2$ is not Lipschitz in our sense.

(2) We will not specify the smallest constant in many situations, and hence c' is also a Lipschitz constant if $c \leq c'$ is the case. However later when we develop a detailed comparison of dynamics by φ and f_t as above, Lipschitz constants which are less than 1 will play an important role to induce some uniform bounds of orbit rates.

Lemma 3.3. *Any relative* $(\max, +)$*-function is Lipschitz.*

Proof. This holds since relative $(\max, +)$-functions are piecewisely linear. $\qquad \square$

Let us introduce the notation:

$$P_N(c) = \begin{cases} \frac{c^{N-n+1}-1}{c-1} & c \neq 1, \\ (N - n + 1) & c = 1. \end{cases}$$

$P_N(c)$ is monotone increasing with respect to N for any $c \geq 0$. Notice that the equalities hold:

$$cP_N(c) + 1 = P_{N+1}(c).$$

$P_N(c)$ grows:

$$\begin{cases} \text{exponentially} & c > 1 \\ \text{linearly} & c = 1 \\ \text{uniformly bounded} & c < 1 \end{cases}$$

Let $\{\varphi, \varphi_t, f_t\}$ be a tropical triplet (see the presentations $*_A, *_B, *_C$ in 2.3), and M be the number of the components in 2.2.

Lemma 3.4. *The uniform estimates:*

$$|\varphi(x_0, \ldots, x_{n-1}) - \varphi_t(x_0, \ldots, x_{n-1})| \leq \log_t M$$

hold for any $(x_0, \ldots, x_{n-1}) \in \mathbb{R}^n$.

Proof. We claim that the uniform estimate:

$$|a_1 \oplus_t \cdots \oplus_t a_m - \max(a_1, \ldots, a_m)| \leq \log_t m$$

holds. One may assume $\max(a_1, \ldots, a_m) = a_1$ without loss of generality. Then we have the estimates:

$$|a_1 \oplus_t \cdots \oplus_t a_m - \max(a_1, \ldots, a_m)| = |\log_t(t^{a_1} + \cdots + t^{a_m}) - a_1|$$
$$= |\log_t[t^{a_1}(1 + t^{a_2 - a_1} + \cdots + t^{a_m - a_1})] - a_1|$$
$$= |a_1 + \log_t(1 + t^{a_2 - a_1} + \cdots + t^{a_m - a_1}) - a_1|$$
$$= |\log_t(1 + t^{a_2 - a_1} + \cdots + t^{a_m - a_1})| \leq \log_t m$$

where the last estimate holds since $a_i - a_1 \leq 0$ and so $t^{a_i - a_1} \leq 1$ holds for $t > 1$.

Now we have the estimates:

$$|(a_1 \oplus_t \cdots \oplus_t a_m - b_1 \oplus_t \cdots \oplus_t b_l) - (\max(a_1, \ldots, a_m) - \max(b_1, \ldots, b_l))|$$
$$\leq |a_1 \oplus_t \cdots \oplus_t a_m - \max(a_1, \ldots, a_m)| + |b_1 \oplus_t \cdots \oplus_t b_l - \max(b_1, \ldots, b_l)|$$
$$\leq \log_t m + \log_t l = \log_t(ml) = \log_t M.$$

$\square$

Let us denote the orbits:

$$x_{N+1} = \varphi(x_{N-n+1}, \ldots, x_N),$$
$$x'_{N+1} = \varphi_t(x'_{N-n+1}, \ldots, x'_N)$$

with the same initial values $\bar{x}_0 = \bar{x}'_0 \in \mathbb{R}^n$.

Lemma 3.5. *Let $c \geq 0$ and M be the Lipschitz constant and the number of the components for φ respectively. Then the estimates hold:*

$$|x_N - x'_N| \leq P_N(c) \log_t M.$$

Proof. Let us denote $\bar{x}_N = (x_N, \ldots, x_{N+n-1}) \in \mathbb{R}^n$. Then $\bar{x}_0$ is the initial value and $x_{N+n} = \varphi(\bar{x}_N)$ hold for all $N \geq 0$. Similarly we denote $\bar{x}'_N = (x'_N, \ldots, x'_{N+n-1}) \in \mathbb{R}^n$ with $x'_{N+n} = \varphi_t(\bar{x}'_N)$.

Firstly the estimate $|x'_n - x_n| \leq \log_t M$ holds by lemma 3.4, since $x_n = \varphi(x_0, \ldots, x_{n-1})$ and $x'_n = \varphi_t(x_0, \ldots, x_{n-1})$.

Since φ is c-Lipschitz and $\bar{x}_1 - \bar{x}'_1 = (0, \ldots, 0, x_n - x'_n)$, the estimates:

$$|x_{n+1} - x'_{n+1}| = |\varphi(\bar{x}_1) - \varphi_t(\bar{x}'_1)|$$
$$\leq |\varphi(\bar{x}_1) - \varphi(\bar{x}'_1)| + |\varphi_t(\bar{x}'_1) - \varphi(\bar{x}'_1)|$$
$$\leq c\, d(\bar{x}_1, \bar{x}'_1) + \log_t M$$
$$= c|x_n - x'_n| + \log_t M \leq (c+1)\log_t M$$

hold. Next we have the estimates:

$$|x_{n+2} - x'_{n+2}| = |\varphi(\bar{x}_2) - \varphi_t(\bar{x}'_2)|$$
$$\le |\varphi(\bar{x}_2) - \varphi(\bar{x}'_2)| + |\varphi(\bar{x}'_2) - \varphi_t(\bar{x}'_2)| \le c\, d(\bar{x}_2, \bar{x}'_2) + \log_t M$$
$$= c\max(|x_{n+1} - x'_{n+1}|, |x_n - x'_n|) + \log_t M \le [c(c+1)+1]\log_t M.$$

Now let us proceed by induction. Suppose the estimates $|x_{n+l} - x'_{n+l}| \le P_{n+l}(c)\log_t M$ hold for all $l \le k-1$. Then we have the estimates:

$$|x_{n+k} - x'_{n+k}| = |\varphi(\bar{x}_k) - \varphi_t(\bar{x}'_k)|$$
$$\le |\varphi(\bar{x}_k) - \varphi(\bar{x}'_k)| + |\varphi(\bar{x}'_k) - \varphi_t(\bar{x}'_k)| \le c\, d(\bar{x}_k, \bar{x}'_k) + \log_t M$$
$$= c\max(|x_k - x'_k|, \ldots, |x_{n+k-1} - x'_{n+k-1}|) + \log_t M$$
$$\le (cP_N(c)+1)\log_t M = P_{N+1}(c)\log_t M.$$

This completes the induction step. $\qquad\qquad\qquad\qquad\qquad\square$

3.4 Comparison between rational orbits

Let f_t and g_t be two elementary rational functions, and consider the orbits:

$$z_{m+1} = f_t(z_{m-n+1}, \ldots, z_m), \quad w_{m+1} = g_t(w_{m-n+1}, \ldots, w_m)$$

with the same initial values:

$$(z_0, \ldots, z_{n-1}) = (w_0, \ldots, w_{n-1}) \in \mathbb{R}^n_{>0} = (0, \infty) \times \cdots \times (0, \infty).$$

Let φ be the tropical correspondence to f_t. We denote by M_f and c_f as the number of the components and the Lipschitz constant of φ respectively. Notice that c_f and c_g are the same if f and g are tropically equivalent, since the Lipschitz constant is determined by the function itself rather than its presentation. On the other hand M_f and M_g can be very different from each other.

Theorem 3.1. *The following are equivalent:*

(1) f_t and g_t are tropically equivalent.

(2) Any two orbits with the same initial value satisfy the uniformly bounded rates from each other:

$$\left(\frac{z_N}{w_N}\right)^{\pm 1} \equiv \left\{\frac{z_N}{w_N}, \frac{w_N}{z_N}\right\} \le M^{2P_N(c)}$$

where $c = \max(c_f, c_g)$ and $M = \max(M_f, M_g)$.

(3) There is a function $F : \mathbb{N} \to \mathbb{N}$ which is independent of the initial values and $t > 1$ so that the uniformly bounded rates:

$$\left(\frac{z_N}{w_N}\right)^{\pm 1} \equiv \{\frac{z_N}{w_N}, \frac{w_N}{z_N}\} \le F(N)$$

hold for all $N \ge 0$.

Proof. Let $(\varphi, \varphi_t, f_t)$ and (ψ, ψ_t, g_t) be the tropical triplets. With respect to the same initial values $x_i = y_i = \log_t z_i$ for $0 \le i \le n - 1$, let us denote the orbits by $\{x_N\}_N$ and $\{y_N\}_N$ given by φ and ψ respectively.

Let us put $x'_N = \log_t(z_N)$ and $y'_N = \log_t(w_N)$ respectively. Then by lemma 3.2, $\{x'_N\}_N$ is the orbit for φ_t and $\{y'_N\}_N$ is for ψ_t with the same initial values $x'_i = y'_i = x_i = y_i$ for $0 \le i \le n - 1$.

By lemma 3.5, the estimates hold:

$$|x_N - x'_N|, \ |y_N - y'_N| \ \le P_N(c) \log_t M.$$

Let us verify that (1) implies (2). Suppose f_t and g_t are tropically equivalent, and so φ and ψ are the same functions. Thus $x_N = y_N$ holds, and hence we have the estimates:

$$\log_t \left(\frac{z_N}{w_N}\right)^{\pm} \le |\log_t(z_N) - \log_t(w_N)| = |x'_N - y'_N|$$

$$\le |x_N - x'_N| + |x_N - y_N| + |y_N - y'_N| \le 2P_N(c) \log_t M = \log_t M^{2P_N(c)}.$$

Since $\log_t$ are monotone increasing, we have the estimates:

$$\left(\frac{z_N}{w_N}\right)^{\pm} \le M^{2P_N(c)}.$$

(3) follows immediately from (2).

Let us verify that (3) implies (1). Let us fix an initial value $x_i = y_i = x'_i = y'_i$ for $0 \le i \le n - 1$, and put $z_i = t^{x_i}$. Suppose the uniform estimates $\left(\frac{z_N}{w_N}\right)^{\pm} \le F(N)$ hold. By applying $\log_t$ on both sides, the estimates $|x'_N - y'_N| \le \log_t F(N)$ hold. Then by lemma 3.5, the estimates:

$$|x_N - y_N| \le |x_N - x'_N| + |x'_N - y'_N| + |y'_N - y_N|$$
$$\le 2P_N(c) \log_t M + \log_t F(N)$$

hold for each $N \ge 0$. Now the equalities $x_N = y_N$ must hold for all N, since the left hand side is independent of t, while the right hand side converges to 0 as $t \to \infty$.

$\square$

Example 3.2. If the Lipschitz constant $c > 1$ is larger than 1, then $P_N(c)$ grows exponentially with respect to N. So the ratios between orbits above are bounded at most double exponentially. Such double exponential estimates are optimal as we will see below.

Let us take $l, k \geq 1$, and consider two iterated dynamics:

$$z_N = f(z_{N-1}) = (z_{N-1})^l, \quad w_N = g(w_{N-1}) = 2(w_{N-1})^k.$$

f and g are tropically equivalent if and only if $k = l$.

Let $z_0 = w_0 > 0$ be any initial value. Then a direct calculation gives:

$$z_N = (z_0)^{l^N}, \quad w_N = 2^{\frac{k^N-1}{k-1}}(w_0)^{k^N} = 2^{\frac{k^N-1}{k-1}}(z_0)^{k^N}.$$

Thus if $l = k$, then the equality:

$$\frac{w_N}{z_N} = 2^{\frac{l^N-1}{l-1}}$$

holds, which satisfies the uniformly double exponential bound.

On the other hand if $k > l$, then

$$\frac{w_N}{z_N} = 2^{\frac{k^N-1}{k-1}} z_0^{k^N - l^N}$$

which heavily depends on the initial values.

Let f be an elementary rational function, and take a positive number $a \in \mathbb{R}$. Then af is tropically equivalent to f if $a \in \mathbb{Q}$. On the other hand af is not tropically equivalent to f if $a \in \mathbb{R}\backslash\mathbb{Q}$. Actually it is not even elementary rational.

On the other hand one can have the iterated dynamics by the same way. So the statement (3) in the above theorem makes sense.

Question 3.2. *Can one find an example of f and $a \in \mathbb{R}\backslash\mathbb{Q}$ which breaks the above uniform bounds in (3)?*

3.5 Quasi-recursiveness

Let $(\varphi, \varphi_t, f_t)$ be a tropical triplet. We have already seen the examples of recursive dynamics in 3.1.

Firstly let us consider what happens to φ when f_t is recursive of period M for each $t > 1$. In this case we show that φ also satisfies the same property.

Lemma 3.6. *If f_t is recursive of period p, then φ is also recursive so that p is divisible by the period.*

Proof. Let $\{x_N\}_N$ and $\{x'_N\}_N$ be the orbits for φ and φ_t with the same initial values $(x_0, \ldots, x_{n-1}) = (x'_0, \ldots, x'_{n-1}) \in \mathbb{R}^n$, and put $z_i = t^{x'_i}$ for $i \geq 0$.

It follows from lemma 3.5 that for any small $\epsilon > 0$, there is $t_0 > 1$ so that the estimates $|x_i - x'_i| \leq \epsilon$ hold for any $0 \leq i \leq n - 1 + p$ and $t \geq t_0$.

By assumption, $z_{j+p} = z_j$ holds for $0 \leq j \leq n - 1$, and so $x'_{j+p} = x'_j$ also holds by lemma 3.2. Then we have the estimates:

$$|x_{j+p} - x_j| \leq |x_{j+p} - x'_{j+p}| + |x'_{j+p} - x'_j| + |x'_j - x_j| \leq 2\epsilon.$$

Since the left hand side is independent of t, and $\epsilon > 0$ is arbitrarily small, the equalities $x_{j+p} = x_j$ must hold. Thus φ gives a recursive dynamics of period p' where p is divisible by p'. $\qquad\square$

The converse to lemma 3.6 is not true in general (see example 3.1).

Definition 3.2. Let f_t be an elementary rational function. It is quasi-recursive of period p bounded by a constant $C \geq 1$, if for any $t > 1$ and any initial value $(z_0, \ldots, z_{n-1}) \in \mathbb{R}_{>0}^n$, the orbits $\{z_N\}_N$ for f_t satisfy the uniform estimates for all $N \geq 0$:

$$\left(\frac{z_{N+p}}{z_N} \right)^{\pm} \equiv \left\{ \frac{z_{N+p}}{z_N}, \frac{z_N}{z_{N+p}} \right\} \leq C.$$

We will assume that p is minimal unless otherwise stated.

Remark 3.4. Recursivity is equivalent to $C = 1$, since at least one of $\frac{z_{N+p}}{z_N}$ or $\frac{z_N}{z_{N+p}}$ is larger than or equal to 1.

Let us characterize quasi-recursive rational dynamics by the corresponding recursive (max, +)-dynamics.

Let $(\varphi, \varphi_t, f_t)$ be a tropical triplet, and M and c be the number of the components and the Lipschitz constant respectively.

Theorem 3.2. *The following are equivalent:*

(1) φ is recursive of period p.

(2) There is a constant $C \geq 1$ such that f_t is quasi-recursive of period p bounded by C.

(3) f_t is quasi-recursive of period p bounded by $M^{2P_{n-1+p}(c)}$.

Proof. Let $\{x_N\}_N$ and $\{x'_N\}_N$ be the orbits for φ and φ_t respectively, where we choose the initial values $x_i = x'_i = \log_t z_i$ for $0 \le i \le n-1$.

From (1) to (3): It is enough to verify the conclusion for $0 \le N \le n-1$. $|x_N - x'_N| \le P_N(c) \log_t M$ holds by lemma 3.5. $x_{N+p} = x_N$ holds since φ is recursive. Then the estimates hold:

$$\left(\frac{z_N}{z_{N+p}}\right)^{\pm} = \left(\frac{t^{x'_N}}{t^{x'_{N+p}}}\right)^{\pm} \le t^{|x'_N - x'_{N+p}|} \le t^{2P_{N+p}(c)\log_t M} = M^{2P_{N+p}(c)}$$

where we used the estimates:

$$|x'_{N+p} - x'_N| \le |x'_{N+p} - x_{N+p}| + |x_{N+p} - x_N| + |x_N - x'_N|$$
$$\le 2P_{N+p}(c) \log_t M.$$

Thus one can put $C = M^{2P_{n-1+p}(c)}$.

(3) implies (2) trivially.

From (2) to (1): We have the estimates:

$$|x_{N+p} - x_N| \le |x_{N+p} - x'_{N+p}| + |x'_{N+p} - x'_N| + |x'_N - x_N|$$
$$\le 2P_N(c) \log_t M + |\log_t \frac{z_N}{z_{N+p}}| \le 2P_N(c) \log_t M + \log_t C$$

for all $0 \le N \le n-1$. Since the left hand side is independent of $t > 1$, (1) follows by letting $t \to \infty$.

Notice that the above proof verifies that they have the same minimum period p. $\square$

Corollary 3.1. *Quasi-recursivity is preserved in tropically equivalent class.*

Proof. This follows from theorem 3.2 since recursivity does not depend on the choice of presentations over relative $(\max, +)$-functions. $\square$

Let us introduce an invariant of quasi-recursive dynamics. Let f_t be an elementary rational function and $[f_t]$ be the tropically equivalent class.

Then the quasi-recursive constant of the class with the period p is given by:

$$C([f_t], p) = \inf_{[g_t]=[f_t]} \sup_{t>1}$$
$$\sup_{(w_0,\dots,w_{n-1})\in\mathbb{R}^n_{>0}} \{(\frac{w_{N+p}}{w_N})^{\pm} : w_{N+1} = g_t(w_{N-n+1},\dots,w_N)\}.$$

Properties:

(1) If f_t is quasi-recursive of period p, then the bound:

$$C([f_t], p) \leq M^{2P_{n-1+p}(c)}$$

holds by theorem 3.2.

Moreover $C(f_t, p')$ are all infinity for any $1 \leq p' \leq p - 1$.

(2) $C([f_t], p) = 1$ holds, if there exists a recursive representative in the tropically equivalent class.

(3) Suppose f is quasi-recursive. Then af is also the case for any $a \in \mathbb{Q}$.

Question 3.3.

(4) *Suppose f is quasi-recursive. Can one find some $a \in \mathbb{R}_{>0}$ so that af is not quasi-recursive?*

(5) *Does the converse statement to (2) still hold? It would be interesting to analyze whether the infimum can be achieved by an elementary rational function.*

Let us look at additivity of quasi-recursive maps.

Corollary 3.2. *Let f_t and f'_t be two elementary rational functions. Assume that:*

(1) the corresponding (max, +)*-functions φ and φ' satisfy the pointwise estimates $\varphi \geq \varphi'$.*

(2) φ is recursive.

Then $f_t + f'_t$ is quasi-recursive.

Proof. $f_t + f'_t$ admits $\max(\varphi, \varphi')$ as the corresponding (max, +)-function. On the other hand $\max(\varphi, \varphi') = \varphi$ holds as functions by the inequality $\varphi \geq \varphi'$. In particular the left hand side is also recursive. Thus the conclusion follows by theorem 3.2. $\qquad\qquad\square$

Example 3.3. Let $f(z, w) = \frac{1 + w^2}{zw}$. The corresponding (max, +)-function is given by $\varphi(x, y) = \max(-y, y) - x$, which is recursive of period 9 (see example 3.1). So f is quasi-recursive of period 9.

Let $f'(z, w) = z^{-1}$ and $\varphi'(x, y) = -x$. Then clearly the pointwise estimate $\varphi \geq \varphi'$ holds, and so by corollary 3.2

$$(f + f')(z, w) = \frac{1 + w + w^2}{zw}$$

is also quasi-recursive. One can see that both f and $f + f'$ are not recursive by computer calculation. By the same argument, $(f + lf')(z, w) = \frac{1 + lw + w^2}{zw}$ are all quasi-recursive for $l = 0, 1, \ldots$

References

The study on iteration dynamics has a long history. [MS] is a comprehensive textbook on one dimensional iteration dynamics. The comparison method is based on Lipschitz analysis on piecewisely linear maps [Kat4]. Lipschitz property is fundamental in the analysis of differential equations or dynamical systems. See introductory textbooks [Log], [Nit] on such analysis. See [GKP] for concrete examples of recursive dynamics.

Chapter 4

Rational perturbation of dynamics in two variables

4.1 Stationary dynamics

Tropical equivalence gives a kind of deformation among elementary rational functions. It is somehow 'discrete' deformation such as f to $2f$. One can also consider continuous deformation among them, and in this chapter we present a continuous deformation theory of real rational dynamics whose behavior can be understood passing through tropical transform.

Recall that one can extend the class of elementary rational functions to allow functions of the form:

$$f_t(\bar{z}) \equiv \frac{k_t(\bar{z})}{h_t(\bar{z})} = (\ \Sigma_{k=1}^m \ t^{\alpha_k} \ \bar{z}^{\bar{a}_k} \ )(\ \Sigma_{k=1}^l \ t^{\beta_k} \ \bar{z}^{\bar{b}_k} \ )^{-1}$$

with $\bar{a}_k, \bar{b}_k \in \mathbb{R}^n$ (see remark 2.1(3)). We also call such functions as elementary rational.

Let f_t be an elementary rational function, and consider its continuous deformation f_t^ϵ with $f_t^0 = f_t$ in the class for $-\epsilon_0 < \epsilon < \epsilon_0$. We say that it is an ϵ-perturbation of f_t.

Let us say that:

(1) the iterated dynamics given by f_t is *quasi-stable*, if any orbit is bounded away from both zero and infinity.

(2) It is *quasi-unstable* if there is a constant C independent of t so that for any initial value $\bar{z}_0$ with $|\bar{z}_0| \geq C$, the orbits are unbounded.

Definition 4.1. An ϵ-perturbation of f_t is stationary, if it is quasi-stable for $\epsilon > 0$ and quasi-unstable for $\epsilon < 0$ for all sufficiently small $|\epsilon| << 1$.

The aim in this chapter is to verify:

Theorem 4.1. *(1) There is an elementary rational function f_t of 2 variables which admits a stationary perturbation.*

(2) The same property holds if we change the ϵ-perturbation of f_t to another tropically equivalent family g_t of perturbation.

Explicitly we will choose:

$$f_t^\epsilon(z, w) = z^\epsilon \frac{1 + w}{z}.$$

On the other hand we develop a general perturbation theory extensively.

4.2 Contraction vs boundedness

Let $\psi(x_0, \ldots, x_{n-1})$ be a relative $(\max, +)$-function. Recall that it has the following form where $\bar{a}_i, \bar{b}_j \in \mathbb{R}^n$:

$$\psi(\bar{x}) = \max(\alpha_1 + \bar{a}_1\bar{x}, \ldots, \alpha_m + \bar{a}_m\bar{x}) - \max(\beta_1 + \bar{b}_1\bar{x}, \ldots \beta_l + \bar{b}_l\bar{x}).$$

Let us regard it as a self-map for $\bar{x}_0 = (x_0, \ldots, x_{n-1})$:

$$\psi : \mathbb{R}^n \to \mathbb{R}^n$$
$$\psi(\bar{x}_0) = (x_1, \ldots, x_{n-1}, \psi(\bar{x}_0)).$$

As in chapter 3, we denote $x_N = \psi(x_{N-n}, \ldots, x_{N-1})$ inductively. Then we have the formulas:

$$\psi(\bar{x}_0) = (x_1, \ldots, x_{n-1}, \psi(\bar{x}_0)) = (x_1, \ldots, x_n),$$
$$(\psi)^2(\bar{x}_0) = (x_2, \ldots, x_{n-1}, \psi(\bar{x}_0), \psi(x_1, \ldots, x_{n-1}, \psi(\bar{x}_0))) = (x_2, \ldots, x_{n+1})$$
$$(\psi)^3(\bar{x}_0) = (x_3, \ldots, x_{n-1}, x_n, x_{n+1}, \psi(x_2, \ldots, x_{n+1})) = (x_3, \ldots, x_{n+2})$$

$\ldots$

$$(\psi)^l(\bar{x}_0) = (x_l, \ldots, x_{n+l-2}, \psi(x_{l-1}, \ldots, x_{n+l-2})) = (x_l, \ldots, x_{n+l-1})$$

$\ldots$

Recall the norm on $\mathbb{R}^n$ in sec. 3.3.

Definition 4.2. ψ is contracting, if there are $0 < \mu < 1$ and $l \geq 1$ so that the estimate:

$$|\varphi(x_0, \ldots, x_{n-1})| \leq (1 - \mu)|(x_0, \ldots, x_{n-1})|$$

holds for $(x_0, \ldots, x_{n-1}) \in \mathbb{R}^n$, where $\varphi = \psi^l$.

Remark 4.1. Since we regard a function as a map between $\mathbb{R}^n$ as above, l should be at least n to give the contracting estimate as above.

Let f_t be the corresponding elementary rational function to ψ.

Proposition 4.1. *Suppose ψ is contracting. Then:*

(1) f_t is quasi-stable. More strongly there is some constant C independent of $t > 1$ and initial values so that for any orbit $\{z_N\}_N$ of f_t, there is some N_0 and for all $N \geq N_0$, the estimates hold:

$$C^{-1} \leq z_N \leq C.$$

(2) For any other g_t which is tropically equivalent to f_t, it also satisfies the above property (1).

Proof. Step 1: (2) follows from (1) since contracting property does not depend on the choice of presentations of relative $(\max, +)$-functions.

Step 2: Let us verify (1), and let (ψ, ψ_t, f_t) be the tropical triplet. For any initial value $(z_0, \ldots, z_{n-1}) \in \mathbb{R}^n_{>0}$, let:

$$\{z_N\}_N, \quad \{x_N\}_N, \quad \{x'_N\}_N$$

be the orbits for f_t, ψ and ψ_t respectively, where $x_i = x'_i = \log_t z_i$ for $0 \leq i \leq n-1$. Then $x'_N = \log_t z_N$ holds for all $N \geq 0$ (see chapter 3).

Let us also regard $\psi_t : \mathbb{R}^n \to \mathbb{R}^n$ by $\psi_t(\bar{x}_0) = (x_1, \ldots, x_{n-1}, \psi_t(\bar{x}_0))$, and $\varphi_t = \psi_t^l$ the same way as $\varphi = \psi^l$.

Let us denote the components of φ by $(\varphi^0, \ldots, \varphi^{n-1})$. It follows from the above computation that the formulas hold:

$$x_{l+i} = \varphi^i(x_0, \ldots, x_n)$$

for $0 \leq i \leq n-1$. The equalities:

$$\varphi_t^i(x_0, \ldots, x_{n-1}) = x'_{l+i}$$

also hold where $\varphi_t = (\varphi_t^0, \ldots, \varphi_t^{n-1})$ for $0 \leq i \leq n-1$.

φ^i are given by some compositions by ψ, which are also relative $(\max, +)$-functions, since the class is preserved under compositions.

Step 3: Let $M \geq 1$ be the maximum number of the components among φ^i. Then the estimates hold by lemma 3.4:

$$|\varphi_t(x_0, \ldots, x_{n-1}) - \varphi(x_0, \ldots, x_{n-1})| \leq \log_t M.$$

Let us choose a small $\mu' > 0$ so that $\epsilon \equiv \mu - \mu' > 0$ is still positive. Then put the unique number α with $\log_t M = \alpha\mu'$ or $t = M^{\frac{1}{\alpha\mu'}}$ so that we obtain the estimate:

$$|\varphi_t(x_0, \ldots, x_{n-1})| \leq (1 - \mu)|(x_0, \ldots, x_{n-1})| + \alpha\mu'.$$

Let $\bar{p}_N = (\varphi_t)^N(x_0, \ldots, x_{n-1}) \in \mathbb{R}^n$. Suppose $\bar{p}_N$ satisfy $|\bar{p}_N| > \alpha$ for some $N \geq 0$. Then the above estimate gives the inequality:

$$|\varphi_t(\bar{p}_N)| \leq (1 - \epsilon)|\bar{p}_N|.$$

We claim that there is some $N_0 \geq 0$ and a constant C independent of $t > 1$ so that all points $\bar{p}_N$ satisfy uniform bounds:

$$|\bar{p}_N| \leq C\alpha$$

for all $N \geq N_0$. By the above inequality, it is enough to verify this when $|\bar{p}_N| \leq \alpha$ holds.

Then each component of φ_t is of the form:

$$\log_t(t^{L_1^1} + \cdots + t^{L_{m_1}^1}) - \log_t(t^{L_1^2} + \cdots + t^{L_{m_2}^2})$$

where L_j^i are linear functions. Let:

$$\max(L_1^i(\bar{p}_N), \ldots, L_{m_i}^i(\bar{p}_N)) = L_{l_i}^i(\bar{p}_N)$$

for $i = 1, 2$. Then the estimate holds:

$$|\varphi_t(\bar{p}_N)| \leq |L_{l_1}^1(\bar{p}_N)| + |L_{l_2}^2(\bar{p}_N)| + \log_t m_1 + \log_t m_2$$
$$\leq |L_{l_1}^1(\bar{p}_N)| + |L_{l_2}^2(\bar{p}_N)| + \log_t M \leq C|\bar{p}_N| + \log_t M$$
$$\leq C|\bar{p}_N| + \mu'\alpha \leq (C + \mu')\alpha$$

where C can be twice the maximum of the Lipschitz constants of the linear functions L_j^i. This verifies the claim.

Step 4: Notice the equalities $p_N^i = x'_{Nl+i}$ hold for $\bar{p}_N = (p_N^0, \ldots, p_N^{n-1})$. Since the equality $z_N = t^{x'_N} = M^{\frac{x'_N}{\alpha\mu'}}$ holds, the orbits:

$$M^{\frac{-C}{\mu'}} \leq z_{Nl+i} \leq M^{\frac{C}{\mu'}}, \quad N \geq N_0, \quad (0 \leq i \leq n - 1)$$

are ultimately bounded both from below and above, which are independent of $t > 1$.

If $l \leq n$, then we are done.

Suppose $l > n$. Then we regard $(z_n, \ldots, z_{2n-1})$ as the initial value, and apply the above estimates. Then:

$$M^{\frac{-C}{\mu'}} \leq z_{Nl+i+n} \leq M^{\frac{C}{\mu'}}$$

holds for $0 \leq i \leq n - 1$ and all $N \geq N_1 >> 0$. We iterate the same argument for the initial values $(z_{kn}, \ldots, z_{(k+1)n-1})$, $k = 0, 1, \ldots, s = [\frac{l}{n}]$, and obtain the same bounds for all $N \geq N_k >> 0$. Finally we put $L = \max(N_0, \ldots, N_s)$. Then the conclusion follows for all z_N with $N \geq L$.

$\square$

Remark 4.2. The estimates above show that in small neighborhoods of 1, some fluctuation of orbits $\{z_N\}_N$ occurs. They give us no information on the behavior of the orbits near 1, and they will not converge to 1 in general. While at infinity the orbits $\{x_N\}_N$ are contracting.

For later purpose, let us generalize proposition 4.1. Let ψ be a $(\max, +)$-function. Let us say that ψ is:

(1) *eventually contracting*, if there is some $0 < \mu < 1$ and m_0, l so that for any initial value $(x_0, \ldots, x_{n-1}) \in \mathbb{R}^n$, there is some $m \leq m_0$ and the estimates hold for all $i = 0, 1, 2, \ldots$:

$$|\psi^{li}(\psi_t^m((x_0, \ldots, x_{n-1})))| \leq (1 - \mu)^i |\psi_t^m(x_0, \ldots, x_{n-1})|.$$

(2) *essentially contracting*, if there is some $0 < \mu < 1$ and m_0, n_0 so that for any initial value $(x_0, \ldots, x_{n-1}) \in \mathbb{R}^n$, there are $m \leq m_0$ and indices $l_0 < l_1 < l_2, \cdots \to \infty$, $l_{i+1} - l_i \leq n_0$, such that the estimates hold:

$$|\psi^{l_i}(\psi_t^m(x_0, \ldots, x_{n-1}))| \leq (1 - \mu)^i |\psi_t^m(x_0, \ldots, x_{n-1})|$$

for all $i = 0, 1, 2, \ldots$ We will say that it is essentially contracting with respect to (μ, m_0, n_0). Eventually contracting implies essentially contracting.

Corollary 4.1. *Suppose ψ is essentially contracting. Then f_t is quasistable. More strongly there is some constant C independent of $t > 1$ and initial values, so that for any orbit $\{z_N\}_N$, there is some N_0 and for all $N \geq N_0$, the estimates hold:*

$$C^{-1} \leq z_N \leq C.$$

Proof. For the eventually contracting case, one can choose the initial value $\psi_t^m(x_0, \ldots, x_{n-1})$ rather than $(x_0, \ldots, x_{n-1})$. Then the conclusion follows by proposition 4.1.

Let us consider the essentially contracting case. Again one may replace the initial value by $\psi_t^m(x_0, \ldots, x_{n-1})$. Let $\{x_N\}_N$, $\{x_N'\}_N$ and $\{z_N\}_N$ be orbits for ψ, ψ_t and f_t respectively. Then by assumption, there are indices $l_1 < l_2 < \cdots \to \infty$, $l_{i+1} - l_i \leq n_0$, and $|\psi^{l_i}(x_0, \ldots, x_{n-1})| \leq (1 - \mu)^i |(x_0, \ldots, x_{n-1})|$ holds. Let us denote $\bar{x}_N' = (x_N', \ldots, x_{N+n-1}')$.

We follow a similar argument as the proof of proposition 4.1, and give a sketch of the proof below.

Let us denote $\psi^l = (\psi_0^l, \ldots, \psi_{n-1}^l)$ and denote m_l^i as the number of the components of ψ_i^l. Then put $o = \max\{m_l^i : 0 \leq i \leq n-1, 0 \leq l \leq n_0 + 1\}$, and $m = o^n$. As before we put $\log_t m = \alpha\mu'$. Then there is a constant C

and indices $N_0 < N_1 < \ldots, N_{i+1} - N_i \leq n_0 + 1$, so that $|\bar{x}_{N_i}| \leq C\alpha$ holds for all i. Since $N_{i+1} - N_i$ are uniformly bounded, it follows by replacing the constant by a larger $C' \geq C$ if necessary, that $|\bar{x}'_N| \leq C'\alpha$ holds for all $N \geq N_0$. Thus uniformity $C^{-1} \leq z_N \leq C$ holds for all $N \geq N_0$.

□

4.3 Unbounded orbits

Let φ be a relative $(\max, +)$-function on $\mathbb{R}^n$. We say that it is *homogeneous*, if $\varphi(\alpha x_0, \ldots, \alpha x_{n-1}) = \alpha\varphi(x_0, \ldots, x_{n-1})$ holds for all $\alpha \in \mathbb{R}$ and $(x_0, \ldots, x_{n-1}) \in \mathbb{R}^n$.

Let us take initial values $\bar{x}_0 = (x_0, \ldots, x_{n-1}) \in \mathbb{R}^n$ and denote the orbits by x_N defined by φ. Then for a homogeneous φ, we put:

$$L(N) = \inf_{|\bar{x}_0|=1} \sup_{l \leq N}\{x_l; \ x_{n+i} = \varphi(x_i, \ldots, x_{n+i-1})\}.$$

Let us say that the dynamics by φ is *positively unbounded*, if there is some N_0 such that $L(N_0) > 1$ holds.

Let f_t be the corresponding elementary rational function, and consider the orbits by the dynamics $z_{n+i} = f_t(z_i, \ldots, z_{n+i-1})$.

Lemma 4.1. *Suppose φ is positively unbounded. Then for any g_t tropically equivalent to f_t, it is quasi-unstable.*

In fact there is a constant C so that g_t has unbounded orbits for any initial value $\bar{z}_0$ with $|\bar{z}_0| \geq M^C$, where M is the number of the components of g_t.

Proof. Let us denote the $(\max, +)$-function corresponding to g_t by ψ. ψ is also homogeneous and positively unbounded. By assumption, there is some N_0 such that $L(N_0) > 1 + \delta$ holds for some $\delta > 0$. Let us put the number of the components of g_t by $M = M_{g_t}$. Let $\{x_N\}_N$ and $\{x'_N\}_N$ be the orbits of ψ and ψ_t with the same initial values $\bar{x}_0 = (x_0, \ldots, x_{n-1}) \in \mathbb{R}^n$ respectively. Then there is some constant C independent of choice of g_t so that the corresponding orbit satisfies the estimates $|x_i - x'_i| < C \log_t M$ for $0 \leq i \leq N_0$ by lemma 3.5.

Now choose any initial value $\bar{x}_0$ with $\frac{\delta}{2}|\bar{x}_0| > C \log_t M$. By assumption, there is some $n_0 \leq N_0$ so that $x_{n_0} \geq |\bar{x}_0|(1 + \delta)$ holds. Then we have the estimates:

$$x'_{n_0} > |\bar{x}_0|(1 + \delta) - \log_t M^C > |\bar{x}_0|\left(1 + \frac{\delta}{2}\right).$$

Now we choose another initial value $\bar{x}_{0,1} = (x'_{n_0}, \ldots, x'_{n_0+n-1})$, and denote the orbits by $x'_{N,1}$ and $x_{N,1}$ for ψ_t and ψ respectively for $N = 0, 1, \ldots$. Then since ψ is homogeneous, there is $0 \leq n_1 \leq N_0$ so that $x_{n_1,1} \geq |\bar{x}_0|(1+\delta)(1+\frac{\delta}{2})$ holds. Thus the estimate:

$$x'_{n_1,1} > |\bar{x}_0|(1+\delta)\left(1+\frac{\delta}{2}\right) - \log_t M^C > |\bar{x}_0|(1+\delta)$$

holds. Notice $x'_{n_1,1} = x'_{n_0+n_1}$ lies on the orbits of ψ_t with the initial value $(x_0, \ldots, x_{n-1})$.

Let us iterate the same process. Let $\bar{x}_{0,2} = (x'_{n_1,1}, \ldots, x'_{n_1+n-1,1})$ be another initial value, and denote the orbits as $x'_{N,2}$ and $x_{N,2}$ for ψ_t and ψ respectively, $N = 0, 1, \ldots$. Then there is $0 \leq n_2 \leq N_0$ so that $x_{n_2,2} \geq |\bar{x}_0|(1+\delta)(1+\delta)$, and so $x'_{n_2,2} > |\bar{x}_0|(1+\delta)^2 - \log_t M^C > |\bar{x}_0|(1+\frac{3}{2}\delta)$ holds, where $x'_{n_2,2} = x'_{n_0+n_1+n_2}$. The same process gives $x'_{n_k,k}$ with the estimates:

$$x'_{n_k,k} > |\bar{x}_0|\left(1+\frac{(k+1)\delta}{2}\right)$$

for $k = 0, 1, \ldots$, and $x'_{n_k,k} = x'_{\sum_{i=0}^{k} n_i}$.

Now all points $x'_{n_k,k}$ lie on the orbits $\{x'_n\}_n$ by ψ_t with the initial value $(x_0, \ldots, x_{n-1})$. Let us put $z_N = t^{x'_N}$. Then the bound $|\bar{x}_0| > \log_t M^{C'}$ holds if $z_i > M^{C'}$ is satisfied for some $0 \leq i \leq n-1$, where $C' = \frac{2C}{\delta}$. Since $\{z_N\}_N$ is the orbit for g_t, the result follows.

$\square$

Remark 4.3. Even if φ is homogeneous, the corresponding f_t is not necessarily t independent. For example consider $\varphi(x) = \max(2x, -2x, -1+x)$.

4.4 Traces

Let φ be a relative (max, +)-function in two variables. Let $\bar{x}_0 = (x_0, x_1) \in \mathbb{R}^2$ be an initial value, and denote the orbit of φ by $\{x_N\}_N$. Let us plot the sequence of points $(x_N, x_{N-1}) \in \mathbb{R}^2$, and regard φ as a map:

$$\varphi : (x_{N-1}, x_N) \to (x_N, x_{N+1}) \quad N \geq 1.$$

Let us denote $\varphi^i(x_0, x_1) = \bar{x}_i = (x_i, x_{i+1})$ for $i = 0, 1, \ldots$

Let L_0 be the straight line which connects $\bar{x}_0$ and $\bar{x}_1$ in $\mathbb{R}^2$. The *trace* of φ with the initial value $\bar{x}_0$ is a connected piecewise linear line L in $\mathbb{R}^2$:

$$L = \cup_{i \geq 0} \varphi^i(L_0) \subset \mathbb{R}^2.$$

L contains all the points $\cup_{i\geq 0}\,\bar{x}_i \subset L$.

For later purpose let us describe traces more explicitly. For simplicity of the notation, let φ be of the form $\varphi(x,y) = \max(\alpha_1(x,y),\ldots,\alpha_m(x,y))$, where α_j are linear functions.

Suppose $\varphi(\bar{x}_0) = \alpha_j(\bar{x}_0)$ for some $1 \leq j \leq m$. If the equality $\varphi(\bar{x}) = \alpha_j(\bar{x})$ holds for any point $\bar{x} \in L_0$, then $\varphi(L_0)$ is a straight line.

Otherwise, let $l_0^1 \subset L_0$ be a subline with one end point $\bar{x}_0$ so that the equality $\varphi(\bar{x}) = \alpha_j(\bar{x})$ holds for any point $\bar{x} \in l_0^1$. We choose l_0^1 of maximal length with this property, and denote the other end point by $\bar{x}_0^1$. We put $j = j_0$.

Then there is another $j_1 \neq j_0$ so that $\varphi(\bar{x}_0^1) = \alpha_{j_1}(\bar{x}_0^1)$ holds. Let $\bar{x}_0^2$ be another point on L_0 so that the equality $\varphi(\bar{x}) = \alpha_{j_1}(\bar{x})$ holds for any point $\bar{x} \in l_0^2$, where l_0^2 is a subline on $L_0 \backslash l_0^1$ which connects $\bar{x}_0^1$ and $\bar{x}_0^2$. Again we choose l_0^2 of maximal length with this property. Then one finds another $j_2 \neq j_1$ so that $\varphi(\bar{x}_0^2) = \alpha_{j_2}(\bar{x}_0^2)$ holds, and we seek for $\bar{x}_0^3$ the same way. By iterating this process, finally one divides L_0 into smaller sublines $L_0 = l_0^1 \cup l_0^2 \cup \cdots \cup l_0^k$ for some k, where one end point of l_0^k is $\bar{x}_1$. By the construction, the images $l_1^j \equiv \varphi(l_0^j)$ are all straight lines. Then we have a broken line as a union of line segments:

$$L_0 \cup \varphi(L_0) = \cup_{i=0}^k l_0^i \cup_{i=0}^k l_1^i.$$

We do the same process, by replacing the role of L_0 by l_1^i, of dividing all l_1^i into smaller sublines $l_1^i = \cup_{j=0}^{k'} l_1^{i,j}$. Then again the images $l_2^{i,j} \equiv \varphi(l_1^{i,j})$ are all straight lines, and we have a broken line as a union of line segments:

$$L_0 \cup \varphi(L_0) \cup \varphi^2(L_0) = \cup_i l_0^i \cup_{i,j} l_1^{i,j} \cup l_2^{i,j}.$$

This way one obtains piecewisely linear lines L which we call the *trace* by continuing this process possibly infinitely many times.

For the general case $\varphi = \max(\alpha_1,\ldots,\alpha_m) - \max(\beta_1,\ldots,\beta_l)$, one can similarly obtain the trace L of φ as a broken line. In fact there is some $1 \leq j \leq m$, $1 \leq j' \leq l$ and a subline $l_0^1 \subset L_0$ so that the restriction of φ on l_0^1 satisfies $\varphi(\bar{x}) = \alpha_j(\bar{x}) - \beta_{j'}(\bar{x})$ which is linear, and repeat it the same way.

Example 4.1. Suppose φ is recursive of period p. Then the trace L consists of l-gon with $l \geq p$. In particular it consists of a finite number of segments.

4.5 Return map for trace

Let L be the trace for φ with an initial point $(0,0) \neq \bar{x} \in \mathbb{R}^2$, and $l_{\bar{x}}$ be the half infinite straight line containing the origin and $\bar{x}$. The *return map* for L is an assignment:

$$r(\bar{x}) \in l_{\bar{x}} \cap L$$

where on the connected subline $C \subset L$ between $\bar{x}$ and $r(\bar{x})$, $C \cap l_{\bar{x}}$ consists of only these two end points. Of course it can happen that $r(\bar{x}) = \phi$ in some cases.

$r^2(\bar{x}) \in C' \cap L$ is another assignment, where on the connected subline $C' \subset L$ between $r(\bar{x})$ and $r^2(\bar{x})$, $C' \cap l_{\bar{x}}$ consists of only these two end points. $r^k(\bar{x})$ is similar.

Remark 4.4. (1) By the construction, $r^l(\bar{x}) \neq \phi$ for all $1 \leq l \leq k$, if $r^k(\bar{x}) \neq \phi$.

(2) Later we choose $l_{\bar{x}}$ as the x-axis $[0,\infty) \times \{0\}$.

(3) Throughout the rest of chapter 4, we always assume that φ is homogeneous.

Let L be the trace of φ with the initial point $(1,0)$, and suppose $r^k(1,0) \neq \phi$ for some $k \in \{1,2,\dots\}$. Then there is some $n_0 \geq 1$ so that $r^k(1,0) \in \varphi^{n_0}(L_0)$. For two points $\bar{x}, \bar{y} \in L$, let us denote by $l(\bar{x},\bar{y}) \subset L$ the broken segment in L which connects these two points.

Let s_k be the straight line connecting $(1,0)$ and $r^k(1,0)$ on the x-axis, and denote $S_k = l((1,0), r^k(1,0)) \cup s_k \subset \mathbb{R}^2$. It is the image of S^1 by the piecewisely linear map into $\mathbb{R}^2$. Let us say that S_k is *non trivial*, if it does not contain the origin and is not contractible in $\mathbb{R}^2 \setminus \{(0,0)\}$.

Now $\varphi^{n_0}(L_0)$ splits as two broken segments:

$$l(\varphi^{n_0}(1,0), r^k(1,0)) \cup l(r^k(1,0), \varphi^{n_0+1}(1,0))$$

and denote the connected broken line:

$$C_k = l(r^k(1,0), \varphi^{n_0+1}(1,0)) \cup \varphi(l(\varphi^{n_0}(1,0), r^k(1,0))) \subset L.$$

Lemma 4.2. *Suppose $r_k \neq \phi$ for some k. Then for any $\bar{x} \in L_0$, there is some $\bar{x}' \in C_k$ and $a > 0$ so that $\bar{x} = a\bar{x}'$.*

Proof. The end points of C_k is $r^k(1,0) = (b,0)$ and $\varphi(r^k(1,0)) = \varphi(b,0)$ for some $b > 0$. Thus it is a broken line connecting $(b,0)$ and $\varphi(b,0)$. Since L_0 is a segment connecting $(1,0)$ and $\varphi(1,0)$, and since φ is homogeneous, thus the conclusion follows. $\qquad\square$

We say that φ is *focusing*, if there is some $r_k \neq \phi$ so that for any point $\bar{x} \in C_k$, $\bar{x} \in aL_0$ hold for some $a > 0$.

Example 4.2. All homogeneous and recursive maps are focusing.

Let us introduce the real numbers:
$$D_i^k(\varphi) = \inf\{|\bar{y}| - |\bar{x}| : \bar{x} \in C_k, \ \bar{y} \in L_0\},$$
$$D_s^k(\varphi) = \sup\{|\bar{y}| - |\bar{x}| : \bar{x} \in C_k, \ \bar{y} \in L_0\}.$$

We call the *degree* of φ as the minimum $k \geq 1$ such that
(1) φ is focusing with respect to k,
(2) S_k is non trivial, and
(3) one of $D_i^k \geq 0$ or $D_s^k \leq 0$ holds.
If there is no such k, then the degree of φ is 0.

Example 4.3. (1) $\varphi(x, y) = \max(0, y) - x$ is recursive of period 5, and its degree is 1.
(2) $\varphi(x, y) = \max(-y, y) - x$ is recursive of period 9, and its degree is 2.

Definition 4.3. Let φ be focusing of degree $k \geq 1$.
(1) It is stably focusing, if the inequality $D_i^k(\varphi) > 0$ is satisfied.
(2) It is unstably focusing, if $D_s^k(\varphi) < 0$ is satisfied.

Let φ be a homogeneously relative $(\max, +)$-function and f_t be the corresponding elementary rational function. Now we consider the tropical correspondence of dynamics:

Proposition 4.2. *Suppose φ is stably focusing. Then f_t is quasi-stable. More strongly there is some constant C independent of $t > 1$ and initial values, so that for any orbit $\{z_N\}_N$, there is some N_0 and for all $N \geq N_0$, the estimates hold:*
$$C^{-1} \leq z_N \leq C.$$

Proof. Let $k \geq 1$ be the degree of φ, and C_k and n_0 be as above.
Let (z_0, z_1) be any initial value, and put $x_0 = \log_t z_0, x_1 = \log_t z_1$. Then we have orbits $\{x_N\}_N, \{x'_N\}_N, \{z_N\}_N$ for φ, φ_t, f_t respectively.
By lemma 3.5, there is some constant C so that $|x_i - x'_i| \leq \log_t M^C$ holds for all $0 \leq i \leq n_0 + 1$, where M is the number of the components of f_t.

It follows from homogeneity of φ and the above estimate that if the initial value $\bar{x}_0$ satisfies the estimate $|\bar{x}_0| \geq c \log_t M^C$ for large $c \gg 1$, then there is some $a > 0$, $0 \leq m \leq n_0 + 1$ so that $\bar{y}_0 = \varphi_t^m(\bar{x}_0)$ lies in the $\log_t M^C$ neighborhood of aL_0, where aL_0 is a scale change of L_0 by a.

Now we claim that if $|\bar{x}_0| \geq c \log_t M^C$ holds, then there is a constant $1 > \mu > 0$ and some n_0 so that φ is essentially contracting with respect to $(\mu, n_0 + 1, n_0 + 2)$.

On the other hand if $\max(z_i, z_i^{-1}) \geq M^{Cc}$ holds for one of $i = 0, 1$, then the estimate $|\bar{x}_0| \geq c \log_t M^C$ follows. Combining with these, the conclusion follows by corollary 4.1.

Let us verify the claim. Assume $|\bar{x}_0| \geq c \log_t M^C$ holds for large c and choose m and $\bar{y}_0$ as above. Then for $n_0 - 1 \leq p \leq n_0 + 2$, $\bar{y} = \varphi^p(\bar{y}_0) \in aC_k$. Since we have chosen a large c, it follows from homogeneity that there is some $0 < \mu < 1$ so that $|\bar{y}| \leq (1 - \mu)|\bar{y}_0|$ holds.

Next by assumption, $\bar{y} \in cL_0$ for some $c > 0$. Then it follows again by homogeneity that for $p' = n_0$ or $n_0 + 1$, $\varphi^{p'}(\bar{y}) = \varphi^{p+p'}(\bar{y}_0) \in cC_k$, and the estimate $|\varphi^{p+p'}(\bar{y}_0)| \leq (1 - \mu)|\varphi^p(\bar{y}_0)| \leq (1 - \mu)^2|\bar{y}_0|$ holds.

One can iterate the second step to see that φ is essentially contracting. This verifies the claim. $\qquad\square$

Remark 4.5. When φ is recursive and hence not contracting, then the above proof gives no information on boundedness of orbits for the corresponding rational function. Actually one needs deeper analysis to obtain any effective information on f_t even if it admits an ϵ-stationary perturbation.

4.6 Perturbation of recursive maps

Let φ be a relative $(\max, +)$-function of two variables, and regard it as a map $\varphi : \mathbb{R}^2 \to \mathbb{R}^2$. A *conserved polygon* is a polygon $P \subset \mathbb{R}^2$ so that it contains the orbit (x_{n-1}, x_n) with any initial value $(x_0, x_1) \in P$.

Example 4.4. (1) When φ is recursive, then closed traces are conserved polygons.

(2) If φ is homogeneous, then they are scale invariant, in the sense that if P is a conserved polygon, then rP is also the same for any $r > 0$.

Let $|\epsilon_i| \ll 1$ be two small numbers with $i = 0, 1$. ϵ_i *perturbation* of φ is given by:

$$\varphi^i(x_0, x_1) = \varphi(x_0, x_1) + \epsilon_i x_i.$$

Let $f_t(z_0, z_1)$ be the elementary rational function. Correspondingly ϵ_i perturbation of f_t is given by:

$$f_t^{\epsilon_i}(z_0, z_1) = z_i^{\epsilon_i} f_t(z_0, z_1).$$

Let $(x_0, x_1) \in \mathbb{R}^2$ be an initial value, and $\{x_n\}_n$ be the orbits for φ^i. Throughout chapter 4, we regard φ^i as maps $\varphi^i : (x_{n-1}, x_n) \to (x_n, x_{n+1})$, and study distributions of the sequences of points $(x_{n-1}, x_n) \in \mathbb{R}^2$.

4.7 A recursive map

Let us calculate an example. Let:

$$z_{N+1} = f(z_{N-1}, z_N) = \frac{1 + z_N}{z_{N-1}}$$

be the recursive map of period 5 (see example 3.1(1)). Its tropical correspondence is given by:

$$x_{N+1} = \varphi(x_{N-1}, x_N) = \max(0, x_N) - x_{N-1}$$

whose orbits also have period 5. Here we study dynamics given by ϵ_0 perturbation of the homogeneously relative (max, +)-function:

$$\varphi^\epsilon(x_0, x_1) = \max(0, x_1) - x_0 + \epsilon x_0.$$

Let $(x_0, x_1) = (1, 0)$ be the initial value. Then the orbit of φ is given by:

$$(1, 0), (0, -1), (-1, 0), (0, 1), (1, 1)$$

and the pentagon P_5 with the vertices above is mapped by φ into itself. Thus P_5 is a conserved pentagon, and since the equation is homogeneous, any rP_5 is also the case for any $r \geq 0$.

Theorem 4.2. *Let g_t be tropically equivalent to $f(z, w) = \frac{1+w}{z}$. Then the ϵ_0 perturbation $g_t^\epsilon(z, w) = z^\epsilon g_t(z, w)$ is stationary.*

Example 4.5. $g^\epsilon(z, w) = z^\epsilon(\frac{1+w}{z} + l\frac{1}{z})$, $l = 0, 1, \ldots$ are all stationary.

Thus we have to verify the following properties:

(1) If $\epsilon > 0$ is positive, then any orbit $\{z_N\}_N$ of g_t^ϵ is bounded away from both zero and infinity.

(2) If $\epsilon < 0$ is negative, then any orbit of g_t^ϵ whose initial value has large norm is unbounded.

Proof of theorem 4.2. Combining lemma 4.1 and proposition 4.2 with the following lemma verifies theorem 4.2. Let $|\epsilon| << 1$ be sufficiently small. Let L^ϵ be the traces of φ^ϵ with the end point $(1, 0)$ in 4.4, and recall $C_1 \subset L^\epsilon$ in 4.5. Then we have:

Lemma 4.3. φ^ϵ *is focusing, and the degrees of* φ^ϵ *are all equal to one.*

(1) If $\epsilon > 0$, then φ^ϵ is stably focusing with $r(1, 0) < 1$, and C_1 all lie in the interior of P_5 except $(1, 0)$.

(2) If $\epsilon < 0$, then it is unstably focusing with $r(1, 0) > 1$, and C_1 all lie in the exterior of P_5 except $(1, 0)$.

Proof. For any sign of ϵ, $\varphi^\epsilon(1, 0) = (0, -1 + \epsilon)$ holds. Let l_ϵ be the segment connecting the points $(1, 0)$ and $(0, -1 + \epsilon)$, and consider the trace L^ϵ with the initial point $(1, 0)$.

The direct calculation gives its orbit as:

$$p_0 = (1, 0), p_1 = (0, -1 + \epsilon), \ p_2 = (-1 + \epsilon, 0), \ p_3 = (0, (1 - \epsilon)^2),$$
$$p_4 = ((1 - \epsilon)^2, (1 - \epsilon)^2), \ p_5 = ((1 - \epsilon)^2, \epsilon(1 - \epsilon)^2)$$

for both cases of $\pm\epsilon > 0$.

p_6 depends on the signs of ϵ. We have the following:

$$p_6 = \begin{cases} (\epsilon(1 - \epsilon)^2, (-1 + 2\epsilon)(1 - \epsilon)^2) & \epsilon > 0 \\ (\epsilon(1 - \epsilon)^2, -(1 - \epsilon)^3) & \epsilon < 0. \end{cases}$$

Let us introduce S_ϵ which are the broken lines described below. They will differ with respect to the signs of ϵ. Recall $l(p_i, p_{i+1}) \subset L^\epsilon$ in 4.5.

For $\epsilon \geq 0$, $r^1(1, 0)$ is the intersection between $l(p_5, p_6)$ and the x-axis, and so $r^1(1, 0) = ((1 - \epsilon)^3, 0)$. S_ϵ is a broken line connecting the eight points $\{p_0, \ldots, p_6, \varphi^\epsilon(r^1(1, 0)) = (0, -(1 - \epsilon)^4)\}$.

For $\epsilon < 0$, $r^1(1, 0)$ is the intersection between $l(p_4, p_5)$ and the x-axis, and so $r^1(1, 0) = ((1 - \epsilon)^2, 0)$. S_ϵ is a broken line connecting the seven points $\{p_0, \ldots, p_5, \varphi^\epsilon(r^1(1, 0)) = (0, -(1 - \epsilon)^3)\}$.

In both cases, we claim the inclusions $C_1 \subset S_\epsilon \subset L^\epsilon$. Then it is immediate by drawing S_ϵ on the plane to see that it is inside P_5 for $\epsilon > 0$ and outside for $\epsilon < 0$, and the conclusions follow.

Now let us verify the claim. Recall L_0 which is a segment connecting the initial value and its image by φ.

Notice that the bending points are $\varphi^\epsilon(x, y)$ with $y = 0$. Firstly for any point on L_0^ϵ, the y component on the line is non positive. Thus $\varphi^\epsilon(L_0^\epsilon)$ is a straight line, $L_0^\epsilon = l_0^0$ and so $L_0^\epsilon \cup \varphi^\epsilon(L_0^\epsilon) = l_0^0 \cup l_1^0$ in sec. 4.4.

For any point on $l_1^0 = \varphi^\epsilon(L_0^\epsilon)$, the y component on the line is also non positive. Thus $(\varphi^\epsilon)^2(L_0^\epsilon) = l_2^{0,0}$ is a straight line, where $l_1^0 = l_1^{0,0}$.

Next for any point on $l_2^{0,0} = (\varphi^\epsilon)^2(L_0^\epsilon)$, the y component on the lines is non negative, and hence $(\varphi^\epsilon)^3(L_0^\epsilon)$ is a straight line.

By the same reasoning, $(\varphi^\epsilon)^i(L_0)$ are all straight lines for $0 \le i \le 5$ with $\epsilon \ge 0$, and for $0 \le i \le 4$ with $\epsilon < 0$.

Now for $\epsilon \ge 0$, $l(p_6, \varphi^\epsilon(r^1(1,0))) = \varphi^\epsilon(p_5, r^1(1,0))$ are straight lines, and for $\epsilon < 0$, $l(p_5, \varphi^\epsilon(r^1(1,0))) = \varphi^\epsilon(p_4, r^1(1,0))$ are also the same.

These imply that S_ϵ is given by the union of segments, $(\varphi^\epsilon)^i(L_0)$ ($0 \le i \le 5$) with $l(p_6, \varphi^\epsilon(r^1(1,0)))$ for $\epsilon \ge 0$, and $(\varphi^\epsilon)^i(L_0)$ ($0 \le i \le 4$) with $l(p_5, \varphi^\epsilon(r^1(1,0)))$ for $\epsilon < 0$. Thus the inclusion $S_\epsilon \subset L^\epsilon$ holds for both cases, and so $C_1 \subset S_\epsilon$ also holds.   □

References

Recursive maps in two variables have been studied in [GKP], [QRT]. (see [GKP] for more examples). Stationality under rational perturbation comes from an application of tropical geometry to the contraction principle in dynamical systems [Kat 4]. See [Nit] on contraction principle as a basic analytic tool on infinite dimensional dynamical systems, and [Wol] on classification of cellular automata.

Chapter 5

Pentagram map and tropical geometry

The pentagram map consists of a beautiful integrable system, which was originally initiated by R. Schwartz from the viewpoint of the dynamical system on the set of convex polygons in projective geometry. Given an n-gon on the plane, the result of the action called the *pentagram map*, is also another n-gon given by the convex hull of the intersection points of consecutive shortest diagonals.

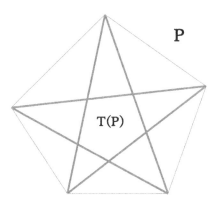

Fig. 5.1 A pentagram map

These polygons will shrink to a point after many compositions, on the other hand many interesting phenomena appear if one considers the induced action on the set of convex polygons modulo projective transformations.

From the viewpoint of the moduli space, it turns out to be quite natural to extend the notion of polygons to *twisted polygons*. The moduli space of twisted n-gons modulo projective transformations constitutes a complete feature in integrable systems. It is equipped with the canonical Poisson structure, Casimirs and the integrals with the Hamiltonian, which commute with the pentagram map. Moreover on the open subset consisting of the universally twisted n-gons, the level set of the Hamiltonian is compact so that the Liouville–Arnold–Jost theory can be applied.

From the viewpoint of scale transform, the continuous limit of the pentagram map T turns out to be the *Boussinesq equation* which plays an important role in integrable systems. In the opposite direction to the scaling limit, we apply tropical geometry.

5.1   Some elements in projective geometry

Let us consider a two dimensional real projective space:

$$\mathbb{R}P^2 = \{(x,y,z) \in \mathbb{R}^3 \setminus \{(0,0,0)\}\} \sim$$

where $(x,y,z) \sim \alpha(x,y,z)$ for any $\alpha \neq 0$.

An element in $\mathbb{R}P^2$ can be interpreted as a line through the origin in $\mathbb{R}^3$. One can canonically embed:

$$\{(x,y,1) \in \mathbb{R}^3\} \cong \mathbb{R}^2 \subset \mathbb{R}P^2$$

by assigning the line through two points $(x,y,1)$ and the origin.

$\mathbb{R}P^2$ is obtained by compactifying the above embedding, or in other words $\mathbb{R}P^2 \setminus \mathbb{R}^2$ consists of a line at infinity.

The group of 3×3 invertible matrices $Gl_3(\mathbb{R})$ act on $\mathbb{R}^3$, which induces the action of the group of projective transformations on $\mathbb{R}P^2$:

$$PGl_3(\mathbb{R}) = Gl_3(\mathbb{R})/\mathbb{R}^\times \to \text{Aut } \mathbb{R}P^2.$$

Lemma 5.1. *Any four points $v_1, \ldots, v_4 \in \mathbb{R}P^2$ can be transformed on the x-axis in $\mathbb{R}^2$ by a projective transformation.*

Proof. Step 1: Suppose there are at most three elements among $\{v_1, v_2, v_3, v_4\}$ which lie at infinity. One can choose an element in $Gl_3(\mathbb{R})$ which is sufficiently near the identity so that it transforms the set $\{v_1, v_2, v_3, v_4\}$ on $\mathbb{R}^2 \subset \mathbb{R}P^2$.

Step 2: Assume that all elements in $\{v_1, v_2, v_3, v_4\}$ lie at infinity, and denote them as $v_i = [x_i, y_i, 0]$.

There is an element in $Gl_3(\mathbb{R})$ which transforms $(x_i, y_i, 0)$ to some (x_i', y_i', z_i') with $z_i' \neq 0$ for $i = 1, 2, 3$ by step 1. Then by using another projective transform which is much nearer the identity compared to the former one, one can transform v_4 with all $[x_i', y_i', z_i']$ on $\mathbb{R}^2$.

Step 3: Let $D \subset \mathbb{R}^2 \subset \mathbb{R}P^2$ be a closed ball which contains all the four points. The range of the image of D by the compact Lie group $PGl_3(\mathbb{R})$ is closed in $\mathbb{R}P^2$.

For small $\epsilon > 0$, the matrix: $\begin{pmatrix} 1 & 0 & 0 \\ 0 & \epsilon & 0 \\ 0 & 0 & 1 \end{pmatrix}$ transforms any $(x, y, 1)$ suffi-

ciently near the x-axis. By proximity, there exists some element in $PGl_3(\mathbb{R})$ which transforms v_i to the x-axis for all $1 \leq i \leq 4$. $\qquad\square$

Lemma 5.2. *Let $g \in Gl_3(\mathbb{R})$. Suppose g transforms two different points $(x_1, 0, 1), (x_2, 0, 1)$ on the x-axis to $(x_1', 0, t), (x_2', 0, t)$ for some $t \neq 0$.*
Then $g(x, 0, 1) = (ax + b, 0, t)$ holds for some a, b, t.

Proof. It follows from assumption that g must have the following form:

$$\begin{pmatrix} a & * & b \\ 0 & * & 0 \\ 0 & * & t \end{pmatrix}$$

So the conclusion holds. $\qquad\square$

Definition 5.1. For $v_1, \ldots, v_4 \in \mathbb{R}P^2$, the cross ratio is defined by:

$$[v_1, v_2, v_3, v_4] = \frac{(x_1 - x_2)(x_3 - x_4)}{(x_1 - x_3)(x_2 - x_4)} \in \mathbb{R} \cup \{\infty\}$$

where $g(v_i) = [x_i, 0, 1]$ for some $g \in PGl_3(\mathbb{R})$.

It follows from lemma 5.2 that the cross ratio is independent of the choice of g above, and so the numerical number is well defined.

5.2 Pentagram map

The pentagram map is a projectively defined natural transformation defined on both untwisted and twisted polygons.

Given an n-gon P on the plane, assign another n-gon P' by the convex hull of the intersection points of consecutive shortest diagonals of P.

Let $\mathbf{C}_n$ be all the set of n-gons on the plane, and denote its assignment by $T : \mathbf{C}_n \to \mathbf{C}_n$. This map itself is less interesting, since after many compositions, it will shrink to a point.

A transformation that maps lines to lines is called a *projective transformation*. Any projective transformation is given by a regular 3×3 matrix in homogeneous coordinates.

Any projective transformation acts on $\mathbf{C}_n$, which commutes with T. Let us denote:

$$\mathfrak{C}_n = \mathbf{C}_n / \sim$$

where $\sim$ is given by projective transformations.

Definition 5.2. The pentagram map is given as:

$$T : \mathfrak{C}_n \to \mathfrak{C}_n$$

which assigns P to P' as above.

Example 5.1. (1) $T = \mathrm{id}$ for $n = 5$ (i.e. for any pentagon P, there is a projective transformation which transforms P to $T(P)$).

(2) $T^2 = \mathrm{id}$ for $n = 6$ (i.e. there is a projective transformation which transforms P to $T^2(P)$).

See [OST1] and also references in the paper.

A *twisted n-gon* is a map:

$$\phi : \mathbb{Z} \to \mathbb{R}P^2$$

such that it satisfies the equalities:

$$\phi(k + n) = M \circ \phi(k) \qquad k \in \mathbb{Z}$$

for some projective automorphism M called *monodromy*.

$\phi(k)$ correspond to vertices on polygons, and so M should be the identity for an untwisted polygon.

It is quite elementary to check that the pentagram map transforms a twisted n-gon to another one. On $\mathbb{R}^2 \subset \mathbb{R}P^2$ by identifying $(v_1 \ v_2)$ with $(v_1 \ v_2 \ 1)$, the action of M can be generically expressed as:

$$(v_1, v_2) \to (v_1, v_2)g + (a, b)$$

for some $g \in Gl_2(\mathbb{R})$ and $a, b \in \mathbb{R}^2$. Then one can check that a polyhedron on the plane is transformed to another one by such an affine transformation (see Fig. 5.3).

Two twisted n-gons ϕ_1, ϕ_2 are *equivalent*, if there is another projective transformation Ψ such that $\Psi \circ \phi_1 = \phi_2$ holds. Two monodromies satisfy the relation $M_2 = \Psi M_1 \Psi^{-1}$.

Let us say that a twisted n-gon ϕ with monodromy M is *generic*, if non zero conditions:

$$\det(\phi(i), \phi(i+1), \phi(i+2)) \neq 0,$$
$$\det(\phi(i), \phi(i+1), M\phi(i+2)) \neq 0,$$
$$\det(\phi(i), M\phi(i+1), M\phi(i+2)) \neq 0$$

hold for all $i \in \mathbb{Z}$. More precisely the condition states that any lift V_i of $\phi(i)$ on $\mathbb{R}^3$ satisfies $\det(V_i, V_{i+1}, V_{i+2}) \neq 0$. Genericity is invariant under equivalent relation.

Definition 5.3. Let us denote $\mathfrak{P}_n$ by the space of generic twisted n-gons modulo equivalence, and denote the extension of the pentagram map by:

$$T : \mathfrak{P}_n \to \mathfrak{P}_n.$$

The pentagram map is not fully defined on the set of twisted n-gons, but surely it does generically.

5.3 Canonical coordinates

Let us recall the *cross ratio* which is a classical invariant in projective geometry. For $t_1, \ldots, t_4 \in \mathbb{R}P^1$, the cross ratio is given by:

$$[t_1, t_2, t_3, t_4] = \frac{(t_1 - t_2)(t_3 - t_4)}{(t_1 - t_3)(t_2 - t_4)}.$$

It is well known that it is invariant under $PGl_2(\mathbb{R})$ action.

Let ϕ be a twisted n-gon, and put $v_i = \phi(i)$.

Definition 5.4. The canonical coordinate is given by:

$$x_i = [v_{i-2}, v_{i-1}, ((v_{i-2}, v_{i-1}) \cap (v_i, v_{i+1})), ((v_{i-2}, v_{i-1}) \cap (v_{i+1}, v_{i+2}))],$$
$$y_i = [((v_{i-2}, v_{i-1}) \cap (v_{i+1}, v_{i+2})), ((v_{i-1}, v_i) \cap (v_{i+1}, v_{i+2})), v_{i+1}, v_{i+2}].$$

Fig. 5.2

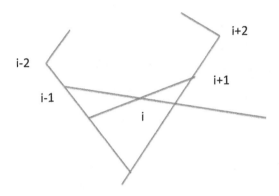

Lemma 5.3. *The assignment:*

$$\Phi : \mathfrak{P}_n \to \mathbb{R}^{2n}$$

given by

$$\phi \to \{(x_1, \ldots, x_n), (y_1, \ldots, y_n)\}$$

is locally homeomorphic.

In particular $\mathfrak{P}_n$ is a smooth manifold of dimension $2n$.

We shall describe the structure of smooth manifolds of $\mathfrak{P}_n$ using another coordinate in sec. 5.4. Actually it is a global coordinate so that these moduli spaces are homeomorphic to the Euclidian space.

Remark 5.1. The cross ratio can be used to describe the Poincaré metric on the upper half plane. Let $z, w \in \mathbb{H}$ and consider the unique semi-circle $C_0 \subset \mathbb{H}$ whose center lies on the real line and on which z, w lie. Let $t, s \in \mathbb{R} \cap C_0$. Then the hyperbolic distance between z and w is given by:

$$\log[z, t, w, s] \geq 0.$$

The main purpose of sec. 5.3 is to verify the following:

Theorem 5.1. *For $i = 1, \ldots, n$, the Pentagram map is given by:*

$$T(z_i, w_i) \equiv (T^*(z_1), \ldots, T^*(z_n), T^*(w_1), \ldots, T^*(w_n))$$

with:

$$\begin{cases} T^*(z_i) = z_i \frac{1-z_{i-1}w_{i-1}}{1-z_{i+1}w_{i+1}}, \\ T^*(w_i) = w_{i+1} \frac{1-z_{i+2}w_{i+2}}{1-z_i w_i}. \end{cases}$$

The proof is not so straightforward, and we pass through another coordinate. This is given in sec. 5.4.

Fig. 5.3 labeling

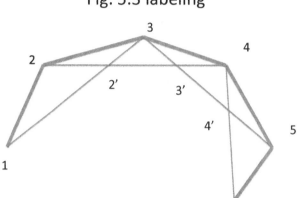

5.4 Another coordinate

Let us introduce another canonical coordinate. Let (a_i) and (b_i) be two n-periodic sequences with $a_{i+n} = a_i$ and $b_{i+n} = b_i$ for all $i \in \mathbb{Z}$. Given such two sequences, consider the orbit $\{V_i\}_{i \in \mathbb{Z}} \subset \mathbb{R}^3$ defined by the difference equations:

$$V_{i+3} = a_i V_{i+2} + b_i V_{i+1} + V_i.$$

Lemma 5.4. *It is deterministic with respect to an initial value (V_0, V_1, V_2). Moreover there is $M \in Sl_3(\mathbb{R})$ with:*

$$V_{i+n} = MV_i.$$

Proof. For $i \geq 3$, V_i are inductively defined by the above formula. For $i \leq -1$, V_i is also determined by the formula:

$$V_i = -b_i V_{i+1} - a_i V_{i+2} + V_{i+3}.$$

As a test case, let us compute V_{i+4} as:

$$\begin{aligned}
V_{i+4} &= a_{i+1} V_{i+3} + b_{i+1} V_{i+2} + V_{i+1} \\
&= a_{i+1}(a_i V_{i+2} + b_i V_{i+1} + V_i) + b_{i+1} V_{i+2} + V_{i+1} \\
&= a_{i+1} a_i V_{i+2} + (a_{i+1} b_i + b_{i+1}) V_{i+1} + a_{i+1} V_i.
\end{aligned}$$

Let us consider the assignment:

$$(V_0, V_1, V_2) \to (V_n, V_{n+1}, V_{n+2}).$$

It follows from the defining equation that this is linear. In fact it gives an isomorphism since the left hand side can be recovered from the right, which implies invertibility. So there is an invertible 3×3 matrix M such that the following holds:

$$\begin{pmatrix} V_n & V_{n+1} & V_{n+2} \end{pmatrix} = M \begin{pmatrix} V_0 & V_1 & V_2 \end{pmatrix}.$$

We claim the equality $\det M = 1$ holds. Let us choose an initial value with $\det(V_0, V_1, V_2) = 1$. Then

$$\begin{aligned}
\det(V_1, V_2, V_3) &= \det(V_1, V_2, a_0 V_2 + b_0 V_1 + V_0) \\
&= \det(V_1, V_2, V_0) = 1
\end{aligned}$$

holds. Inductively, one can verify the equalities $\det(V_i, V_{i+1}, V_{i+2}) = 1$ for all i. Then the above formula implies the claim. $\qquad \square$

Let $\phi = (v_i) \subset \mathbb{R}P^2$ be a twisted n-gon with monodromy $M \in Sl_3(\mathbb{R})$.

Proposition 5.1. *If n is not divisible by 3, then $\mathfrak{P}_n$ admits one to one correspondence with the space of solutions to the above equations. In particular it is homeomorphic to $\mathbb{R}^{2n}$, with the coordinate $\{a_i, b_i\}_{0 \leq i \leq n-1}$.*

Remark 5.2. See [OST1] for information when n is divisible by 3.

Proof. The converse correspondence is given by the above lemma.

Step 1: Let us represent each $M \in PGl_3(\mathbb{R})$ by the unique element $M \in Sl_3(\mathbb{R})$. Notice that both groups are isomorphic.

Let $\phi = (v_i) \subset \mathbb{R}P^2$ be a twisted n-gon with monodromy $M \in Sl_3(\mathbb{R})$ in a general position. Let us choose any lift $\tilde{V}_i \in \mathbb{R}^3$ for $0 \leq i \leq n-1$. The general position guarantees non vanishing $\det(\tilde{V}_i, \tilde{V}_{i+1}, \tilde{V}_{i+2}) \neq 0$. Then let us determine all $\tilde{V}_i$ by the relation $\tilde{V}_{i+n} = M\tilde{V}_i$.

Because of linear independence, there are n-periodic coefficients $a_i, b_i, c_i \in \mathbb{R}$ such that the relations hold:

$$\tilde{V}_{i+3} = a_i \tilde{V}_{i+2} + b_i \tilde{V}_{i+1} + c_i \tilde{V}_i.$$

Let us try to rescale as $\tilde{V}_i = t_i V_i$ so that the equalities hold:

$$\det(V_i, V_{i+1}, V_{i+2}) = 1.$$

If it can be done, then the new relations hold:

$$V_{i+3} = a_i V_{i+2} + b_i V_{i+1} + V_i$$

since the equalities $1 = \det(V_{i+1}, V_{i+2}, V_{i+3}) = c_i \det(V_{i+1}, V_{i+2}, V_i) = c_i$ should hold. Moreover both a_i and b_i are n-periodic.

The conditions are required as:

$$\begin{cases} t_i t_{i+1} t_{i+2} = \det(\tilde{V}_i, \tilde{V}_{i+1}, \tilde{V}_{i+2})^{-1} & (1 \leq i \leq n-2) \\ t_{n-1} t_n t_1 = \det(\tilde{V}_{n-1}, \tilde{V}_n, M\tilde{V}_1)^{-1} \\ t_n t_1 t_2 = \det(\tilde{V}_n, M\tilde{V}_1, M\tilde{V}_2)^{-1}. \end{cases}$$

Let us verify that the system of the equations admit a solution if n is not divisible by 3.

Step 2: Let us reduce the above system of the equations to the case of positive values $\det(\tilde{V}_i, \tilde{V}_{i+1}, \tilde{V}_{i+2}) > 0$ for all i.

One may assume positivity of $\det(\tilde{V}_i, \tilde{V}_{i+1}, \tilde{V}_{i+2})$ for $1 \leq i \leq n-2$. There are three cases where:

$$(\det(\tilde{V}_{n-1}, \ \tilde{V}_n, M\tilde{V}_1), \det(\tilde{V}_n, M\tilde{V}_1, M\tilde{V}_2)) = \begin{cases} (-,+) & (1) \\ (+,-) & (2) \\ (-,-) & (3) \end{cases}$$

where $+$ implies a positive value and $-$ a negative one.

Then replace $\tilde{V}_i$ by $-\tilde{V}_i$ for all i in the following table. For example if (1) happens for $n = 5$, then we change their signs for $i = 2, 3, 5$, and so on.

Step 3: One can choose $\tilde{V}_i$ so that $t_i > 0$ is positive for $1 \leq i \leq n$ by step 2.

By taking log on both sides, it reduces to the system of the equations:

$$\begin{cases} s_i + s_{i+1} + s_{i+2} = \alpha_i & (1 \leq i \leq n-2), \\ s_{n-1} + s_n + s_1 = \alpha_{n-1}, \\ s_n + s_1 + s_2 = \alpha_n. \end{cases}$$

	$n = 3m + 1$	$n = 3m + 2$
(1)	$3l + 1,\ 3l + 2$	$3l\ ,\ 3l + 2$
(2)	$3l,\ 3l + 1$	$3l,\ 3l + 1$
(3)	$3l,\ 3l + 2$	$3l + 1,\ 3l + 2$

It is represented by a matrix:

$$
\begin{pmatrix}
1 & 1 & 1 & 0 & 0 & \ldots & 0 \\
0 & 1 & 1 & 1 & 0 & \ldots & 0 \\
0 & 0 & 1 & 1 & 1 & \ldots & 0 \\
0 & 0 & 0 & 1 & 1 & 1 & \ldots \\
& \vdots & & & & & \\
1 & 0 & \ldots & \ldots & 0 & 1 & 1 \\
1 & 1 & 0 & \ldots & 0 & 0 & 1
\end{pmatrix}
\equiv \begin{pmatrix} \bar{v}_1 & \bar{v}_2 & \bar{v}_3 & \bar{v}_4 & \ldots & \bar{v}_{n-1} & \bar{v}_n \end{pmatrix}
$$

and existence of a solution is guaranteed by invertibility of the matrix.

Suppose there are $a_1, a_2, \ldots, a_n \in \mathbb{R}$ with the equality:

$$a_1 \bar{v}_1 + a_2 \bar{v}_2 + a_3 \bar{v}_3 + \cdots + a_n \bar{v}_n = 0.$$

Then:

$$
\left\{
\begin{array}{l}
a_1 + a_2 + a_3 = 0, \\
\quad \vdots \\
a_{n-2} + a_{n-1} + a_n = 0, \\
a_1 + a_{n-1} + a_n = 0, \\
a_n + a_1 + a_2 = 0.
\end{array}
\right.
$$

It would be instructive to see why the $n = 3m$ case is different from other cases. Suppose $n = 6$. Then

$$
\left\{
\begin{array}{l}
a_1 + a_2 + a_3 = 0, \\
a_2 + a_3 + a_4 = 0, \\
a_3 + a_4 + a_5 = 0, \\
a_4 + a_5 + a_6 = 0, \\
a_5 + a_6 + a_1 = 0, \\
a_6 + a_1 + a_2 = 0.
\end{array}
\right.
$$

In this case let us put $\alpha = a_1 + a_2$. Then the equalities $\alpha = a_2 + a_4 = a_4 + a_5$ holds. Then $a_3 = a_6$ and hence $a_1 + a_5 = \alpha$ follow from the last equality. Now it is clear that the system of the equations admit non trivial solutions.

On the other hand consider the $n = 5$ case:

$$\begin{cases} a_1 + a_2 + a_3 = 0, \\ a_2 + a_3 + a_4 = 0, \\ a_3 + a_4 + a_5 = 0, \\ a_4 + a_5 + a_1 = 0, \\ a_5 + a_1 + a_2 = 0. \end{cases}$$

In this case $\alpha = a_4 + a_5$ and hence $a_3 = a_5 = a_1$ holds. It is immediate then we have the equalities $a_3 = a_2 = a_4$. So all a_i must be the same, and hence $a_i = 0$.

The other cases are parallel to the above two cases.

This assignment is invariant under the projectively equivalent sequences. So this gives an assignment from $\mathbf{P}_n$ to the space of the solutions to the equations above. □

Formulas on vector analysis

Lemma 5.5. *Let* $a, b, c, d \in \mathbb{R}^3$.

(1) $(a \times b) \times (b \times c) = \det(a, b, c)b$

(2) $(a, b) \cap (c, d) = (a \times b) \times (c \times d)$

(3) $[a, b, c, d] = \frac{\lambda_2 \mu_1 - \lambda_1 \mu_2}{\lambda_2 \mu_1}$ *for* $c = \lambda_1 a + \lambda_2 b$ *and* $d = \mu_1 a + \mu_2 b$.

Proof. Notice that length of the vector $a \times b$ is given by the area of the parallelogram spanned by a and b, and the vector $a \times b$ is orthogonal to both a and b.

(1) If three vectors a, b, c are not linearly independent, then both sides vanish. Assume they are linearly independent. Then b is orthogonal to both $a \times b$ and $b \times c$. The length of the left hand side vector surely coincides with the one on the right hand side, since $(a \times b) \times (\frac{b}{||b||} \times c)$ coincides with the volume of the hexagon spanned by a, b, c.

(2) The vector $(a \times b) \times (c \times d)$ lies on the intersection of the two planes spanned by a and b, c and d respectively. Certainly the vector $(a, b) \cap (c, d)$ lies on the line of the intersection.

(3) It is enough to verify the conclusion when both the sums $\lambda_1 + \lambda_2, \mu_1 + \mu_2 \neq 0$ do not vanish. Then since the cross ratio is projectively invariant, one may assume the equalities $\lambda_1 + \lambda_2 = \mu_1 + \mu_2 = 1$.

By abuse of notations, write $a = (a, 0, 1)$, etc., we have the equalities:

$$[a, b, c, d] = \frac{(a-b)\{(\lambda_1 - \mu_1)a + (\lambda_2 - \mu_2)b\}}{\{(1-\lambda_1)a - \lambda_2 b\}\{-\mu_1 a + (1-\mu_2)b\}}$$

$$= \frac{(a-b)\{(1-\lambda_2) - (1-\mu_2)a + (\lambda_2 - \mu_2)b\}}{\lambda_2(a-b)\mu_1(b-a)}$$

$$= \frac{(\lambda_2 - \mu_2)(b-a)}{\lambda_2 \mu_1 (b-a)} = \frac{(\lambda_2 - \mu_2)}{\lambda_2 \mu_1}$$

$$= \frac{(\lambda_2(1-\mu_2) - (1-\lambda_2)\mu_2)}{\lambda_2 \mu_1} = \frac{\lambda_2 \mu_1 - \lambda_1 \mu_2}{\lambda_2 \mu_1}.$$

□

Lemma 5.6. *Assume n is not divisible by 3. Then:*

$$z_i = \frac{a_{i-2}}{b_{i-2}b_{i-1}}, \qquad w_i = -\frac{b_{i-1}}{a_{i-2}a_{i-1}}.$$

Proof. Let us compute the lift of:

$$(v_{i-1}, v_i, ((v_{i-1}, v_i) \cap (v_{i+1}, v_{i+2})), ((v_{i-1}, v_i) \cap v_{i+2}, v_{i+3}))$$

on the plane. Recall:

$$V_{i-1} = V_{i+2} - a_{i-1}V_{i+1} - b_{i-1}V_i.$$

Then we have the equalities by lemma 5.5:

$$(V_{i-1}, V_i) \cap (V_{i+1}, V_{i+2}) = ((V_{i+2} - a_{i-1}V_{i+1} - b_{i-1}V_i) \times V_i) \times (V_{i+1} \times V_{i+2})$$

$$= ((V_{i+2} - a_{i-1}V_{i+1}) \times V_i) \times (V_{i+1} \times V_{i+2})$$

$$= (V_i \times V_{i+2}) \times (V_{i+2} \times V_{i+1}) + a_{i-1}(V_i \times V_{i+1}) \times (V_{i+1} \times V_{i+2})$$

$$= -V_{i+2} + a_{i-1}V_{i+1} \equiv -W_i$$

$$((V_{i-1}, V_i) \cap (V_{i+2}, V_{i+3}))$$

$$= ((V_{i+2} - a_{i-1}V_{i+1} - b_{i-1}V_i) \times V_i) \times (V_{i+2} \times (a_i V_{i+2} + b_i V_{i+1} + V_i))$$

$$= ((V_{i+2} - a_{i-1}V_{i+1}) \times V_i) \times (V_{i+2} \times (b_i V_{i+1} + V_i))$$

$$= b_i(V_{i+2} \times V_i) \times (V_{i+2} \times V_{i+1})$$

$$\quad - a_{i-1}(V_{i+1} \times V_i) \times (V_{i+2} \times V_i) - a_{i-1}b_i(V_{i+1} \times V_i) \times (V_{i+2} \times V_{i+1})$$

$$= b_i V_{i+2} - a_{i-1}V_i - a_{i-1}b_i V_{i+1}.$$

So the lift can be computed as:

$$(W_i - b_{i-1}V_i, \ V_i, \ -W_i, \ b_i W_i - a_{i-1}V_i).$$

For $\alpha = W_i - b_{i-1}V_i$ and $\beta = V_i$, $-W_i = -\alpha - b_{i-1}\beta$ and $b_i W_i - a_{i-1}V_i = b_i\alpha + (b_i b_{i-1} - a_{i-1})\beta$.

By use of the formula (3) in lemma 5.5, we get the result for z_i.

Similar for w_i.

□

Remark 5.3. It can be checked that the right hand side is still well defined, when n is a multiple of 3.

Let $(\mathbb{R}^3)^*$ be a three dimensional Euclidean space which is the dual space to $\mathbb{R}^3$. The *dual projective space* $(\mathbb{R}P^2)^*$ is given by $P((\mathbb{R}^3)^*)$ which consists of the set of lines through the origin in $(\mathbb{R}^3)^*$. Associated to a line in $(\mathbb{R}^3)^*$ is the plane in $\mathbb{R}^3$ which consists of the kernel of the linear map along the line. So an element in $(\mathbb{R}P^2)^*$ is given by a plane in $\mathbb{R}^2$.

Let us introduce two versions of the *projective dualities*:

$$\alpha(v_i), \ \beta(v_i) \ \in \ (\mathbb{R}P^2)^*$$

where $\alpha(v_i)$ is the line (v_i, v_{i+1}), and $\beta(v_i)$ is the line (v_{i-1}, v_{i+1}).

Lemma 5.7.

$$\alpha^2(v_i) = v_{i+1}, \qquad \beta^2(v_i) = v_i, \qquad T = \alpha \circ \beta.$$

Proof. Let us put $\alpha(v_i) = w_i$ and $\beta(v_i) = u_i$, passing through the isomorphism:

$$(\mathbb{R}P^2)^* \cong \mathbb{R}P^2$$

by assigning its orthogonal line to a plane using the Euclidean inner product.

w_i is orthogonal to both v_i and v_{i+1}, and so:

$$\langle w_i, v_i \rangle = \langle w_i, v_{i+1} \rangle = 0, \quad \langle u_i, v_{i-1} \rangle = \langle u_i, v_{i+1} \rangle = 0.$$

$\alpha^2(v_i)$ is orthogonal to w_i and w_{i+1}, whose properties are satisfied for v_{i+1}. So $\alpha^2(v_i) = v_{i+1}$ holds by uniqueness.

Similarly $\beta^2(v_i)$ is orthogonal to u_{i-1} and u_{i+1}, whose properties are satisfied for v_i satisfies such properties, and hence $\beta^2(v_i) = v_i$ holds.

Let us put $p_i = \alpha \circ \beta(v_i)$. p_i has to satisfy the conditions $\langle p_i, u_i \rangle = \langle p_i, u_{i+1} \rangle = 0$. Since $\langle u_{i+1}, v_i - v_{i+2} \rangle = \langle u_i, v_{i-1} - v_{i+1} \rangle = 0$ holds, the intersection of two lines (v_i, v_{i+2}) and (v_{i-1}, v_{i+1}) satisfies the required conditions.

$\qquad\qquad\qquad\qquad\qquad\qquad\qquad\qquad\qquad\qquad\qquad\qquad\qquad\square$

A twisted n-gon (v_i) admits the coordinate $\{a_i, b_i\}_{0 \le i \le n-1}$. Let us denote the coordinate by $\{\alpha^*(a_i), \alpha^*(b_i)\}_i$ which corresponds to $\alpha(v_i)$.

Lemma 5.8. *Assume n is not divisible by 3. Then we have the formulas:*

$$\alpha^*(a_i) = -b_{i+1}, \qquad \alpha^*(b_i) = -a_i.$$

Proof. Let $(V_i) \in \mathbb{R}^3$ be the canonical lift for (v_i) in proposition 5.1. Let us verify that (U_i) is the canonical lift of $\alpha(v_i)$ for $U_i = V_i \times V_{i+1}$.

The triplets (U_i, U_{i+1}, U_{i+2}) are linearly independent, and satisfy:

$$\det(U_i, U_{i+1}, U_{i+2}) = 1.$$

So one can express:

$$U_{i+3} = \alpha^*(a_i)U_{i+2} + \alpha^*(b_i)U_{i+1} + U_i.$$

Consider the equalities:

$$\langle U_{i+1}, V_i \rangle = \langle U_i, V_{i+2} \rangle = 1$$

Then we have the equalities:

$$0 = \langle U_{i+3}, V_{i+3} \rangle = \langle \alpha^*(b_i)U_{i+1} + U_i, V_{i+3} \rangle$$
$$= \alpha^*(b_i)\langle U_{i+1}, V_i \rangle + a_i\langle U_i, V_{i+2} \rangle = \alpha^*(b_i) + a_i.$$

Then the rest of the equalities are obtained as:

$$\alpha^*(a_i) = -\alpha^2(b_i) = -b_{i+1}.$$

$\square$

Lemma 5.9. *The formulas hold:*

$$\beta^*(a_i) = -\frac{\lambda_i b_{i-1}}{\lambda_{i+2}}, \quad \beta^*(b_i) = -\frac{\lambda_{i+3} a_{i+1}}{\lambda_{i+1}}$$

where λ_i is uniquely defined by:

$$\lambda_i \lambda_{i+1} \lambda_{i+2} = -\frac{1}{1 + b_{i-1}a_i}.$$

Proof. The lifts of $\beta(v_i)$ send V_i to:

$$W_i = \lambda_i V_{i-1} \times V_{i+1}$$

where λ_i are chosen so that the equalities $\det(W_i, W_{i+1}, W_{i+2}) = 1$ hold. Let us express:

$$W_{i+3} = \beta^*(a_i)W_{i+2} + \beta^*(b_i)W_{i+1} + W_i$$

and insert the presentation of W_i with the equalities $V_{i+3} = a_i V_{i+2} + b_i V_{i+1} + V_i$. The result is given by:

$$- (\beta^*(a_i)\lambda_{i+2} + b_{i-1}\lambda_i)V_i \times V_{i+1}$$
$$+ (a_{i+1}\lambda_{i+3} + \beta^*(b_i)\lambda_{i+1})V_i \times V_{i+2}$$
$$+ ((1 + b_i a_{i+1})\lambda_{i+3} + \beta^*(a_i)a_i\lambda_{i+2} - \lambda_i)V_{i+1} \times V_{i+2} = 0.$$

Since these three terms are linearly independent, we obtain the expressions for β^*. Then insert the formula for β^* into the last term. The result is:

$$\lambda_{i+3} = \lambda_i \frac{1 + a_i b_{i-1}}{1 + a_{i+1} b_i}.$$

On the other hand, we have the equalities:

$$\begin{aligned}
1 &= \lambda_i \lambda_{i+1} \lambda_{i+2} \det(V_{i-1} \times V_{i+1}, V_i \times V_{i+2}, V_{i+1} \times V_{i+3}) \\
&= -\lambda_i \lambda_{i+1} \lambda_{i+2} (1 + a_i b_{i-1}) \det(V_i \times V_{i+1}, V_{i+1} \times V_{i+2}, V_{i+2} \times V_i) \\
&= -\lambda_i \lambda_{i+1} \lambda_{i+2} (1 + a_i b_{i-1}).
\end{aligned}$$

$\square$

Proof of theorem 5.1

Step 1: Assume that n is not divisible by 3. Then the formulas:

$$\begin{cases}
T^*(a_i) = \alpha^* \circ \beta^*(a_i) = -\beta^*(b_{i+1}) = \frac{\lambda_{i+4} a_{i+2}}{\lambda_{i+2}} \\
T^*(b_i) = \alpha^* \circ \beta^*(b_i) = -\beta^*(a_i) = \frac{\lambda_i b_{i-1}}{\lambda_{i+2}}
\end{cases}$$

hold. So we have the equalities:

$$\begin{aligned}
T^*(z_i) &= \frac{T^*(a_{i-2})}{T^*(b_{i-2}) T^*(b_{i-1})} = \frac{\lambda_{i+2} a_i}{\lambda_i} \frac{\lambda_i}{\lambda_{i-2} b_{i-3}} \frac{\lambda_{i+1}}{\lambda_{i-1} b_{i-2}} \\
&= \frac{a_i}{b_{i-2} b_{i-3}} \frac{1 + a_{i-2} b_{i-3}}{1 + a_i b_{i-1}} = \frac{a_{i-3}}{b_{i-2} b_{i-3}} \frac{1 + a_{i-2} b_{i-3}}{1 + a_i b_{i-1}} \frac{a_i}{a_{i-3}} \\
&= z_{i-1} \frac{1 - (z_{i-1} w_{i-1})^{-1}}{1 - (z_{i+1} w_{i+1})^{-1}} \frac{z_i w_{i-1}}{z_{i+1} w_{i+1}} = z_i \frac{1 - z_{i-1} w_{i-1}}{1 - z_{i+1} w_{i+1}}.
\end{aligned}$$

One can verify the formula for $T^*(w_i)$ in a similar way.

Step 2: We see that the formulas of T are also valid when n is a multiple of 3. Our argument relies on some locality to decide values of $(T^*(z_i), T^*(w_i))$.

Notice the following two properties: (1) to determine the values $T(\phi)(i)$, we need to know the positions of $\phi(i-1), \phi(i), \phi(i+1)$ and $\phi(i+2)$, and (2) the coordinate values (z_i, w_i) are determined by the positions of v_j for $i - 2 \leq j \leq i + 2$.

So to determine the values $(T^*(z_i), T^*(w_i))$, we need to know the positions of v_j for $i - 3 \leq j \leq i + 4$.

Suppose n is a multiple of 3. Let us take a twisted n-gon ϕ, and replace it by another map $\phi' : \mathbb{Z} \to \mathbb{R}P^2$ such that $\phi(i) = \phi'(i)$ for $-2 \leq n + 4$ and extend it so that ϕ' is a twisted $n + 7$-gon. Then by step 1, the desired formulas hold for ϕ', while the values $(T^*(z_i), T^*(w_i))$ coincide for both ϕ and ϕ' for $1 \leq i \leq n$. Hence the formulas also hold for ϕ and $1 \leq i \leq n$. By translating the indices, we know that the desired formulas hold for ϕ and for all $i \in \mathbb{Z}$.

5.5   Integrable Hamiltonian system

In this section, we quickly give an overview of Liouville integrable systems on finite dimensional symplectic manifolds. Then we state complete integrability of the pentagram map. We shall not give proofs for most of the statements, and list some references.

Let (M, ω) be a finite dimensional symplectic manifold, and $H : M \to \mathbb{R}$ be a smooth function. The Hamiltonian vector field X_H is uniquely defined by the pointwise equality:

$$\omega(X_H, \quad) = dH \in T^*M.$$

Example 5.2. Let $\mathbb{R}^{2n} = \mathbb{R}^n \times \mathbb{R}^n$ with the coordinate $(q, p) = (q_1, \ldots, q_n, p_1, \ldots, p_n)$. The standard symplectic form ω_0 is given by:

$$\omega_0 = \Sigma_{i=1}^n dq_i \wedge dp_i.$$

In terms of the Euclidean inner product, it can be expressed as:

$$\omega_0 = \langle \quad, J_0 \quad \rangle, \qquad J_0 = \begin{pmatrix} 0 & I \\ -I & 0 \end{pmatrix}.$$

Let $H : \mathbb{R}^{2n} \to \mathbb{R}$ be a Hamiltonian function. Then the formulas hold:

$$X_H \equiv J_0 \, \mathrm{grad}(H) = \begin{cases} \dot{q}_i = \frac{\partial H}{\partial p_i}, \\ \dot{p}_i = -\frac{\partial H}{\partial q_i}. \end{cases}$$

The latter equalities is the *equation of motion* in classical mechanics. X_H is called the *Hamiltonian vector field* with the Hamiltonian H.

X_H above is uniquely characterized by the pointwise equality:

$$\omega_0(X_H, \quad) = dH.$$

Notice that this formula can be applied to a smooth function over a general symplectic manifold. We also call such a vector field the Hamiltonian vector field.

The following is well known:

Theorem 5.2 (Darboux). *Locally a symplectic manifold of dimension $2n$ is symplectically diffemorphic to an open subset of $(\mathbb{R}^{2n}, \omega_0)$.*

There is a distinguished Poisson bracket over a symplectic manifold. The *Poisson bracket* is a bilinear map:

$$\{ \ , \ \} : C^\infty(M) \times C^\infty(M) \to C^\infty(M)$$

given by:

$$\{G, H\} \equiv \omega(X_G, X_H) = dG(X_H) = -dH(X_G)$$

Lemma 5.10. *The following properties hold:*

(1) $\{f, g\} = -\{g, f\}$,

(2) $\{f, g_1 g_2\} = g_1\{f, g_2\} + g_2\{f, g_1\}$,

(3) $\Sigma_{i=1}^{3}\{\{f_i, f_{i+1}\}, f_{i+2}\} = 0$ (i mod 3).

(4) Let φ_H^t be the flow of X_H. Then

$$\frac{d(G \circ \varphi_H^t)}{dt} = \{G, H\}.$$

In particular H is constant along the flow φ_H^t.

Proof. (1) (2) (3) can all be verified straightforwardly. For (4),

$$\frac{d(G \circ \varphi_H^t)}{dt} = dG(X_H) = \{G, H\}.$$

Since $\frac{d(H \circ \varphi_H^t)}{dt} = 0$ holds, this implies H is constant along the flow φ_H^t.

$\square$

Because of the derivation property (2) above, there is a vector field X_H which satisfies the equality:

$$\{G, H\} = X_H(G) = dG(X_H).$$

There is a pointwise linear map which is called the *Poisson structure*:

$$K : T^*M \to TM$$

by $K(dH) = X_H$.

K is non degenerate since it arises from the symplectic form. Later we will generalize the construction of the Poisson structure over smooth manifolds, and we will see a Poisson structure which is actually degenerate over $\mathbb{R}^{2n}$.

Let H be a Hamiltonian function. G is called an *integral* over H, if $\{H, G\} = 0$ holds. H takes constant values along the flow of X_G.

Definition 5.5. Let (M, ω) be a symplectic manifold of dimension $2n$ with a Hamiltonian function H. The Hamiltonian system (M, ω, H) is called Liouville integrable, if there is a family of functions $F_1, \ldots, F_n$ on M such that:

(1) $\{H, F_i\} = \{F_i, F_j\} = 0$,

(2) $dF_1 \wedge \cdots \wedge dF_n$ does not vanish everywhere on M.

Let (M, ω) be a symplectic manifold. A *Lagrangian submanifold* $L \subset M$ is an n dimensional smooth manifold with $\omega|L \equiv 0$.

Lemma 5.11. *Suppose a Hamiltonian system (M, ω, H) is integrable with $F = (F_1, \ldots, F_n)$. Then*

(1) The level set $M^c = F^{-1}(c)$ gives a foliation of invariant Lagrangian submanifolds.

(2) X_H is tangent to M^c.

Proof. (1) M^c is an n dimensional smooth submanifold by the definition (2) above. Moreover TM_m^c is spanned by $X_{F_i}(m)$ at each point $m \in M^c$. by lemma 5.10. Then the conclusion follows by assumption (1).

(2) follows from lemma 5.10. $\qquad\square$

Let us state the *Liouville–Arnold–Jost theorem*:

Theorem 5.3. *Let (M, ω, H) be an integrable Hamiltonian system.*

If M^c is compact and connected for some $c \in \mathbb{R}^n$, then its neighborhood U is completely foliated into invariant Lagrangian tori.

Moreover there is a symplectomorphism

$$\Phi : T^n \times D \cong U, \qquad \Phi^*(\omega) = \omega_0$$

such that $H \circ \Phi$ is a function which is independent of the coordinate over T^n.

Let us generalize the Poisson structure over finite dimensional smooth manifolds.

Let M be a smooth manifold. A Poisson bracket is a bilinear map:

$$\{ \quad , \quad \} : C^\infty(M) \times C^\infty(M) \to C^\infty(M)$$

which is required to satisfy the following 3 properties:

(1) $\{f, g\} = -\{g, f\}$,

(2) $\{f, g_1 g_2\} = g_1\{f, g_2\} + g_2\{f, g_1\}$,

(3) $\Sigma_{i=1}^{3}\{\{f_i, f_{i+1}\}, f_{i+2}\} = 0$ (i mod 3).

C is called a *Casimir function*, if

$$\{C, \quad\} = 0$$

holds. In the case of symplectic manifolds, the Casimir function is just a constant.

Lemma 5.12. *Let* $\{ \quad , \quad \}$ *be a Poisson structure on a smooth manifold.*

If it is non degenerate so that any Casimir element is a constant, then it defines a symplectic form on M.

More generally let $\bar{C} = (C_1, \ldots, C_m)$ *be a family of Casimir elements. Then generic inverses* $\bar{C}^{-1}(c)$ *consist of symplectic leaves by restriction of the Poisson bracket.*

In the latter case, we say that the Poisson structure has co-rank m. Notice that $n - m$ must be even.

Proof. Let us see that the poisson structure is local in the sense that if a smooth function f vanishes on an open subset $U \subset M$, then $\{g, f\}$ also satisfies the same property. In fact let $\varphi \in C_c^\infty(U)$ and regard it as a smooth function on M. Then we have the equalities:

$$\{g, f\} = \{g, \varphi f\} = \varphi\{g, f\} + f\{g, \varphi\}.$$

Since both φ and f vanish on U, the left hand side is also the same.

This implies that the poisson bracket is a local operator which depends only on df rather than f itself, since $\{g, 1\} = 2\{g, 1\}$ and hence $\{g, \text{const}\} = 0$ holds.

Then for each $f \in C^\infty(M)$, there is a vector field X_f with the equalities $\{g, f\} = X_f(g) = -X_g(f)$. Then define two form ω by the formula $\omega(X_g, X_f) = \{g, f\}$. This gives a non degenerate two form.

The latter case follows from non degeneracy of restriction of the Poisson bracket on each leaf.

$\square$

5.6　Complete integrability of pentagram map

The pentagram map is completely integrable on the moduli space of the twisted n-gons. We shall explain this fact, but we will not give proofs for many of the statements.

Let $\mathfrak{B}_n$ be the moduli space of the generic twisted n-gons with the coordinate $(x_1, \ldots, x_n, y_1, \ldots, y_n)$ in 5.3.

Definition 5.6. The following:

$$\{x_i, x_{i\pm1}\} \pm x_i x_{i+1} = 0, \quad \{y_i, y_{i\pm1}\} = \pm y_i y_{i+1}, \quad \{x_i, y_j\} = 0$$

gives a canonical Poisson bracket:

$$\{\ ,\ \} : C^\infty(\mathbb{R}^{2n}) \times C^\infty(\mathbb{R}^{2n}) \to C^\infty(\mathbb{R}^{2n}).$$

Proposition 5.2. *(1) The Poisson structure is invariant under the pentagram map.*

(2) The Poisson structure on $\mathbf{P}_n$ have co-rank 2 for n odd, and co-rank 4 for n even.

The Casimir functions are under the pentagram map and given by:

$$C_n = \begin{cases} (\ O_n,\ E_n\ ) : \mathbb{R}^{2n} \to \mathbb{R}^2 & \text{for odd } n, \\ (\ O_n,\ E_n,\ O_{n/2},\ E_{n/2}\ ) : \mathbb{R}^{2n} \to \mathbb{R}^4 & \text{for even } n, \end{cases}$$

where:

$$O_n = \Pi_{i=1}^n x_i, \quad E_n = \Pi_{i=1}^n y_i,$$
$$O_{n/2} = \Pi_{2i} x_i + \Pi_{2j+1} x_j, \quad E_{n/2} = \Pi_{2i} y_i + \Pi_{2j+1} y_j.$$

(3) There are other invariants:

$$O_k, \quad E_k \qquad \left(1 \le k \le [\tfrac{n}{2}]\right)$$

which mutually Poisson commute, and are invariant under the pentagram map.

They are algebraically independent for $1 \le k \le [\frac{n}{2}]$ when n is odd and for $1 \le k < [\frac{n}{2}]$ when n is even.

Let us describe how to obtain O_k and E_k. They arise from the monodromy of twisted n-gons, and hence are called the monodromy invariants.

Let $M \in PGl_3(\mathbb{R})$ be the monodromy of a twisted n-gon. Then the quantities:

$$\Omega_1 = \text{trace}^3(M)(\det(M))^{-1}, \quad \Omega_2 = \text{trace}^3(M^{-1})(\det(M^{-1}))^{-1}$$

are both independent of the choice of the lift of M. Then we define:

$$\tilde{\Omega}_1 = O_n^2 E_n \Omega_1, \quad \tilde{\Omega}_2 = O_n E_n^2 \Omega_2$$

which are both functions with variables $(x_1, \ldots, x_n, y_1, \ldots, y_n) \in \mathbb{R}^{2n}$.

Let us introduce a rescaling map:

$$R_t : (x_1, \ldots, x_n, y_1, \ldots, y_n) \to (tx_1, \ldots, tx_n, t^{-1}y_1, \ldots, t^{-1}y_n).$$

The pentagram map commutes with R_t, and the pull back of the above functions admit the formulas:

$$R_t^*(\tilde{\Omega}_1) = \Sigma_{k=1}^{[n/2]} t^k O_k, \quad R_t^*(\tilde{\Omega}_2) = \Sigma_{k=1}^{[n/2]} t^{-k} E_k.$$

So both O_i and E_i are preserved by the pentagram map.

Corollary 5.1. *For generic values c, the symplectic leaves with the monodromy invariants:*

$$(C_n^{-1}(c), O_1, E_1, \ldots, O_{[n/2]}, E_{[n/2]})$$

give the foliations of the Lagrangian submanifolds, which commute with the action by the pentagram map.

The pentagram map is completely integrable on the moduli space of twisted n-gons, with respect to these structures. In order to apply Liouville–Arnold–Jost theorem, $\mathfrak{B}_n$ is still too large to hold such compactness. Let us introduce a smaller but canonical class of their subsets.

Let ϕ be a twisted n-gon with its monodromy $M \in PGl_3(\mathbb{R})$. Let us say that its lift $M \in Sl_3(\mathbb{R})$ is *strongly diagonalizable*, if it has 3 distinct positive real eigenvalues.

The induced action on $\mathbb{R}P^2$ has 3 fixed points with respect to the eigenvectors. Let us connect these three different points by line segments, which constitutes a triangle Δ. The complement of these lines consists of 4 connected components by open triangles, one of which is Δ itself. We call them M-triangles.

Notice that Δ is not unique, since there are two ways to connect two different points by line segments in $\mathbb{R}P^2$. On the other hand because of positivity, Δ is transformed to itself by M.

Let us say that ϕ is *universally convex*, if the following three properties hold:

(1) M is strongly diagonalizable.

(2) $\phi(\mathbb{Z})$ is contained in one of the M-triangles.

(3) The polygonal arc inside the M-triangle obtained by connecting consecutive vertices of $\phi(\mathbb{Z})$ is convex.

Let us introduce another projective invariant. For a twisted n-gon ϕ, the *Hilbert perimeter* is defined by:

$$H = (O_n E_n)^{-1}.$$

Let $\mathbf{U}_n$ be the set of universally convex twisted n-gons modulo projective equivalence.

Lemma 5.13. *(1)* $\mathbf{U}_n$ *is open in* $\mathbf{P}_n$.
(2) $\mathbf{U}_n$ *is invariant under the pentagram map.*
(3) Each level set of H *in* $\mathbf{U}_n$ *is compact.*

Proof. (1) and (2) follow from elementary observation. In particular universal convexity is preserved under a small perturbation of twisted n-gons.

(3) We describe an outline of the proof.

Let $\mathbf{U}_n(M, H)$ be all the set of $\phi \in \mathbf{U}_n$ with monodromy M and the value of the Hilbert perimeter H.

We can normalize the coordinate so that the eigenvectors coincide with the x, y, z axes and the eigenvalues for $(1, 0, 0)$ and $(0, 1, 0)$ satisfy $0 < a < 1 < b$ respectively. Then a universally convex twisted n-gon is contained in the first quadrant.

The equality $H = (\Pi_{i=1}^n m_i)^{-1}$ holds, where:

$$m_k = [(v_i, v_{i-2}), (v_i, v_{i-1}), (v_i, v_{i+1}), (v_i, v_{i+2})]$$

is given by the cross ratio of the slopes. Actually the equalities $m_k = x_k y_k$ hold, which can be verified by straightforward computations.

Let us choose $\phi \in \mathbf{U}_n(M, H)$ and normalize $\phi(0) = (1, 1)$. This is possible since the eigenvalues of $(1, 0)$ and $(0, 1)$ are larger than and smaller than 1 respectively. Then $\phi(n) = (x, y)$ with $x > 1$ and $y < 1$.

Let R be the rectangle whose two opposite corners are $(1, 1)$ and (x, y). Then by convexity, $\phi(0), \ldots, \phi(n)$ are all contained in R.

Suppose the family $\{\phi_k\}_k \subset \mathbf{U}_n(M, H)$ does not admit any convergent subsequence. Then there is some $0 \leq i \leq n$ such that one of the following two things happens: the angle between (v_i, v_{i+1}) and (v_{i+1}, v_{i+2}) tends to 0 but the angle between (v_{i+1}, v_{i+2}) and (v_{i+2}, v_{i+3}) does not as $k \to \infty$, or the distance between v_i and v_{i+1} converges to 0 but v_{i+2} converges to another point.

In the former case $z_{i+2} \to 0$ and $H \to \infty$. In the latter case again $H \to \infty$. This verifies (3).

$\square$

Let us apply the Liouville–Arnold–Jost theorem 5.3:

Theorem 5.4. *Almost every point on $\mathbf{U}_n$ lies on a smooth torus that has a T-invariant affine structure. In particular the orbit of almost every universally convex n-gon undergoes quasi-periodic motion under the pentagram map.*

Proof. Let us verify that the action by T is affine over the invariant tori. By applying theorem 5.3, we obtain the coordinate $(\hat{x}_1, \ldots, \hat{x}_k, \hat{y}_1, \ldots, \hat{y}_k)$ on each generic symplectic leaf for $k = n - 1$ or $n - 2$ when n is odd or even respectively, where the symplectic form is the standard with respect to these coordinates. T preserves the values of $\hat{y}_i$ by proposition 5.2.

Let us denote:

$$T(\bar{x}, \bar{y}) = (\bar{x}', \bar{y}) = (f_1(\bar{x}, \bar{y}), \ldots, f_k(\bar{x}, \bar{y}), \hat{y}_1, \ldots, \hat{y}_k)$$

where $\bar{y} = (\hat{y}_1, \ldots, \hat{y}_k)$. Since T preserves the standard symplectic form, we obtain the equalities:

$$T^*(\omega_0) = \sum_{i=1}^{k} dT^*(x_i) \wedge dy_i$$

$$= \sum_{i,j=1}^{k} \frac{\partial f_i}{\partial x_j} dx_j \wedge dy_i.$$

It follows the equalities:

$$\frac{\partial f_i}{\partial x_j} = \begin{cases} 0 & i \neq j \\ 1 & i = j \end{cases}$$

that there are constants c_i such that:

$$f_i(\bar{x}, \bar{y}) = x_i + c_i$$

holds for all $1 \leq i \leq k$. This implies that T is the translation with respect to this coordinate.

$\square$

5.7 Quasi-recurrence

Let $\mathfrak{B}_n$ be the moduli space of twisted n-gons, and $(\bar{z}_0, \bar{w}_0) \in \mathbb{R}^{2n}$ be the coordinate in theorem 5.1. Let us consider the orbit by the Pentagram map T and denote:

$$T^m(\bar{z}_0, \bar{w}_0) = (z_1^m, \ldots, z_n^m, w_1^m, \ldots, w_n^m)$$

for $m = 0, 1, 2, \ldots$

Theorem 5.5. *There are subsets $U_k \subset \mathbb{R}^{2n}$ so that for any initial value $(\bar{z}_0, \bar{w}_0) \in U_k$, the orbit satisfies the uniform estimates:*

$$1 \leq \max \left(\left| \frac{z_i}{z_i^k} \right|, \left| \frac{z_i}{z_i^k} \right|^{-1}, \left| \frac{w_i}{w_i^k} \right|, \left| \frac{w_i}{w_i^k} \right|^{-1} \right) \leq 4^{(5^k - 1)/4}$$

for all $1 \leq i \leq n$ and $k \geq 0$.

Moreover U_n consists of a non empty open subset in $\mathbb{R}^{2n}$.

Theorem 5.4 does not tell us the sizes of the invariant tori. Combining theorem 5.4 with theorem 5.5 above gives us some uniform bounds of the sizes of these tori as below.

For $\bar{x} \in \mathbf{U}_n \cap U_k$, let $\mathbf{T}_{\bar{x}} \subset \mathbb{R}^{2n}$ be the invariant torus which contains $\bar{x}$. Observe that if the action by T on $\mathbf{T}_{\bar{x}}$ moves $\bar{x}$ a 'small distance', then the orbits $\{\bar{x}, T(\bar{x}), T^2(\bar{x}), \ldots, T^k(\bar{x}), \ldots\}$ will satisfy a property that the distances between $\bar{x}$ and $T^i(\bar{x})$ will increase up to some k.

Regarding this, let us introduce an invariant. For $\bar{x} = (\bar{z}, \bar{w})$, let us denote the orbit by $T^i(\bar{x}) = (z_1^i, \ldots, z_n^i, w_1^i, \ldots, w_n^i)$.

Definition 5.7. We call $k_{\bar{x}} \in \mathbb{N}$ given blow as (T, k)-width:

$$k_{\bar{x}} = \inf \left\{ mk \geq 0 : \max_{1 \leq i \leq n} \left(\left| \frac{z_i}{z_i^{mk}} \right|, \left| \frac{z_i^{mk}}{z_i} \right|, \left| \frac{w_i}{w_i^{mk}} \right|, \left| \frac{w_i^{mk}}{w_i} \right| \right) + 1 \right.$$

$$\left. \geq \sup_a \left\{ \max_{1 \leq i \leq n} \left(\left| \frac{z_i}{z_i^{ak}} \right|, \left| \frac{z_i^{ak}}{z_i} \right|, \left| \frac{w_i}{w_i^{ak}} \right|, \left| \frac{w_i^{ak}}{w_i} \right| \right) \right\} \right\}.$$

Corollary 5.2. *The uniform estimate holds:*

$$1 \leq \sup_{a \geq 0} \max_{1 \leq i \leq n} \left(\left| \frac{z_i}{z_i^{ak}} \right|, \left| \frac{z_i^{ak}}{z_i} \right|, \left| \frac{w_i}{w_i^{ak}} \right|, \left| \frac{w_i^{ak}}{w_i} \right| \right) \leq (\sqrt{2})^{5^{k_{\bar{x}}}}$$

for all $\bar{x} \in \mathbf{U}_n \cap U_k$.

5.7.1   *Proof of theorem 5.5*

The idea of the proof of theorem 5.5 uses tropical geometry analysis developed in chapter 2.

Definition 5.8. The dynamical system $\varphi : \mathbb{R}^{2n} \to \mathbb{R}^{2n}$ given by the formula with $i \mod n$:

$$\varphi \begin{pmatrix} x_i \\ y_i \end{pmatrix} = \begin{pmatrix} x_i + \max(0, x_{i-1} + y_{i-1}) - \max(0, x_{i+1} + y_{i+1}) \\ y_{i+1} + \max(0, x_{i+2} + y_{i+2}) - \max(0, x_i + y_i) \end{pmatrix}$$

is called the pentagram automaton.

Remark 5.4. We will write the dynamics as above for simplicity of the notations, but to be precise it should be denoted such as $\varphi(\{x_i, y_i\}_{i=1}^n)$ and similarly for the output.

For $k \geq 1$, let us denote the set of all the periodic points by:

$$Per_k = \{(\bar{x}, \bar{y}) \in \mathbb{R}^{2n} : \varphi^k(\bar{x}, \bar{y}) = (\bar{x}, \bar{y})\}.$$

Definition 5.9.

$$\mathbf{Per}_k = \{(-t^{x_1}, \ldots, -t^{x_n}, t^{y_1}, \ldots, t^{y_n}) \cup (t^{x_1}, \ldots, t^{x_n}, -t^{y_1}, \ldots, -t^{y_n}) :$$
$$(t^{x_1}, \ldots, t^{x_n}, t^{y_1}, \ldots, t^{y_n}) \in \tilde{Per}_k\}$$

where

$$\tilde{Per}_k = \cup_{t>1} \{(t^{x_1}, \ldots, t^{x_n}, t^{y_1}, \ldots, t^{y_n}) :$$
$$(x_1, \ldots, x_n, y_1, \ldots, y_n) \in Per_k\} \subset \mathbb{R}_{>0}^{2n}$$

for $t > 1$.

Actually we can choose $U_k = \mathbf{Per}_k$. The next lemma verifies that $\mathbf{Per}_n$ is an open subset in $\mathbb{R}^{2n}$.

Lemma 5.14. *Let us put* $\bar{B}_n = \{(\bar{z}, \bar{w}) \in \mathbb{R}_{>0}^{2n} : 0 < z_i w_j \leq 1\}$. *Then we have the inclusion:*

$$\mathbf{D}_n = \{(\bar{z}, \bar{w}) : (\bar{z}, -\bar{w}) \in \bar{B}_n\} \cup \{(\bar{z}, \bar{w}) : (-\bar{z}, \bar{w}) \in \bar{B}_n\} \subset \mathbf{Per}_n.$$

Proof. The inclusion:

$$B_n = \{(\bar{x}, \bar{y}) : x_i + y_j \leq 0\} \subset Per_n$$

is satisfied, since $\varphi^l(x_i, y_i) = (x_i, y_{i+l})$ holds for all l. Then the conclusion holds since the condition $x_i + y_j \leq 0$ is equivalent to $0 < t^{x_i} t^{y_j} \leq 1$. $\square$

Let us apply tropical geometry analysis as in chapter 2. Consider the discrete dynamics:

$$F : \mathbb{R}_{>0}^{2n} \to \mathbb{R}_{>0}^{2n}$$

defined by:

$$F(z_i, w_i) = \left(z_i \frac{1 + z_{i-1} w_{i-1}}{1 + z_{i+1} w_{i+1}}, \ w_{i+1} \frac{1 + z_{i+2} w_{i+2}}{1 + z_i w_i} \right).$$

Let us consider its tropical correspondences:

$$\varphi, \ \varphi_t : \mathbb{R}^{2n} \to \mathbb{R}^{2n}$$

given by:

$$\varphi(\bar{x}, \bar{y}) = \left(\begin{array}{c} x_i + \max(0, x_{i-1} + y_{i-1}) - \max(0, x_{i+1} + y_{i+1}) \\ y_{i+1} + \max(0, x_{i+2} + y_{i+2}) - \max(0, x_i + y_i) \end{array} \right)$$

$$\varphi_t(\bar{x}, \bar{y}) = \left(\begin{array}{c} x_i \otimes_t (0 \oplus_t x_{i-1} \otimes_t y_{i-1}) - 0 \oplus_t x_{i+1} \otimes_t y_{i+1} \\ y_{i+1} \otimes_t (0 \oplus_t x_{i+2} \otimes_t y_{i+2}) - 0 \oplus_t x_i \otimes_t y_i \end{array} \right).$$

Let us denote the orbits with the same initial value by:

$$\varphi^l(x_1, \ldots, x_n, y_1, \ldots, y_n) = (x_{1,l}, \ldots, x_{n,l}, y_{1,l}, \ldots, y_{n,l})$$
$$\varphi_t^l(x_1, \ldots, x_n, y_1, \ldots, y_n) = (x'_{1,l}, \ldots, x'_{n,l}, y'_{1,l}, \ldots, y'_{n,l}).$$

Let $(\bar{z}, \bar{w}) \in \mathbb{R}^{2n}_{>0}$ be an initial value, and consider the orbit:

$$F^l(\bar{z}, \bar{w}) = (\bar{z}_l, \bar{w}_l).$$

Proposition 5.3. *Let $c \geq 1$ and M be the Lipschitz constant and the number of the components for φ respectively. Then the estimates hold:*

$$|x_{i,l+1} - x'_{i,l+1}|, \; |y_{i,l+1} - y'_{i,l+1}| \; \leq \; P_l(c) \log_t M$$

where $P_l(c) = \frac{c^{l+1}-1}{c-1}$.

Proof. For simplicity of the notation, let q imply x or y.

Firstly one has the estimates by lemma 3.4:

$$|q_{i,1} - q'_{i,1}| \leq \log_t M.$$

Since φ is c-Lipschitz, the estimates hold:

$$\begin{aligned} |q_{i,2} - q'_{i,2}| &\leq |\varphi(\bar{q}_1) - \varphi_t(\bar{q}'_1)| \\ &\leq |\varphi(\bar{q}_1) - \varphi(\bar{q}'_1)| + |\varphi_t(\bar{q}'_1) - \varphi(\bar{q}'_1)| \\ &\leq c\, d(\bar{q}_1, \bar{q}'_1) + \log_t M \\ &\leq (c+1) \log_t M \end{aligned}$$

where the metric d was given in chapter 2.

Suppose the conclusion holds up to l. Then we have the estimates:

$$\begin{aligned} |q_{i,l+1} - q'_{i,l+1}| &\leq |\varphi(\bar{q}_l) - \varphi_t(\bar{q}'_l)| \\ &\leq |\varphi(\bar{q}_l) - \varphi(\bar{q}'_l)| + |\varphi_\iota(\bar{q}'_l) - \varphi(\bar{q}'_l)| \\ &\leq c\, d(\bar{q}_l, \bar{q}'_l) + \log_t M \\ &\leq (c P_{l-1}(c) + 1) \log_t M \\ &= P_l(c) \log_t M. \end{aligned}$$

$\square$

Proof of theorem 5.5:

Let us choose an initial value $(\bar{z}, \bar{w}) \in \tilde{Per}_k$ and denote the orbit by:

$$F^l(\bar{z}, \bar{w}) = (\bar{z}_l, \bar{w}_l).$$

Let us choose the initial value as below for $1 \leq i \leq n$:

$$x_i = x_i' = \log_t z_i, \quad y_i = y_i' = \log_t w_i.$$

We denote the orbits $\bar{q}_l = \varphi^l(\bar{q})$ and $\bar{q}_l' = \varphi_t^l(\bar{q})$ respectively, where $\bar{q} = (x_0, \ldots, x_n, y_1, \ldots, y_n)$. By definition one obtains periodicity:

$$\varphi^k(\bar{q}) = \bar{q}.$$

Now we have the estimates by proposition 5.3:

$$d(\bar{q}, \bar{q}_k') \leq d(\bar{q}, \bar{q}_k) + d(\bar{q}_k, \bar{q}_k') = d(\bar{q}_k, \bar{q}_k') \leq \log_t M^{P_{k-1}(c)}.$$

Since we have the equalities:

$$|x_i' - x_{i,k}'| = \log_t(\frac{z_i}{z_i^k})^{\pm}, \quad |y_i' - y_{i,k}'| = \log_t(\frac{w_i}{w_i^k})^{\pm}$$

and since $\log_t$ is monotone, we obtain the estimates:

$$(\frac{z_i}{z_i^k})^{\pm}, \quad (\frac{w_i}{w_i^k})^{\pm} \leq M^{P_{k-1}(c)}.$$

Recall that the Pentagram map is given by:

$$T(z_i, w_i) = \left(z_i \frac{1 - z_{i-1}w_{i-1}}{1 - z_{i+1}w_{i+1}}, w_{i+1} \frac{1 - z_{i+2}w_{i+2}}{1 - z_i w_i}\right).$$

T and F can be transformed by changing the variables:

$$(\bar{z}, \bar{w}) \rightarrow (\bar{z}, -\bar{w}) \text{ or } (-\bar{z}, \bar{w}).$$

In the case of the pentagram map, the number of the component is 4, and the Lipschitz constant is $3 + 2 = 5$.

5.8   Boussinesq equation as a continuous limit

The continuous limit of a twisted n-gon should be a smooth curve:

$$\gamma : \mathbb{R} \rightarrow \mathbb{R}P^2$$

with monodromy $M \in PSl_2(\mathbb{R})$:

$$\gamma(x + 1) = M(\gamma(x)).$$

A curve is *non degenerate*, if $\{\gamma(x), \gamma'(x), \gamma''(x)\}$ are linearly independent at any $x \in \mathbb{R}$, which corresponds to the general position in the discrete case.

Let $\mathfrak{C}$ be the set of non degenerate curves up to projective transformations, which corresponds to $\mathfrak{P}_n$.

Lemma 5.15. *There is one to one correspondence between $\mathfrak{C}$ and the space of the linear differential operators:*

$$A = \left(\frac{d}{dx}\right)^3 + u(x)\frac{d}{dx} + v(x)$$

whose coefficients u and v are both periodic functions.

Proof. Let γ be a non degenerate curve, and choose the unique lift

$$\Gamma : \mathbb{R} \to \mathbb{R}^3$$

with $\det(\Gamma(x), \Gamma'(x), \Gamma''(x)) = 1$. By differentiating, one obtains:

$$
\begin{aligned}
0 &= \det(\Gamma'(x), \Gamma'(x), \Gamma''(x)) \\
&\quad + \det(\Gamma(x), \Gamma''(x), \Gamma''(x)) + \det(\Gamma(x), \Gamma'(x), \Gamma'''(x)) \\
&= \det(\Gamma(x), \Gamma'(x), \Gamma'''(x))
\end{aligned}
$$

so that there are periodic functions u and v such that:

$$\Gamma'''(x) + u(x)\Gamma'(x) + v(x)\Gamma(x) = 0$$

holds. Two non degenerate curves have the same u and v if and only if they are projectively equivalent.

The converse also follows from the above argument.

$\square$

Let γ be a non degenerate curve. For small $\epsilon > 0$, we obtain a new curve γ_ϵ whose tangent vectors at x are parallel to the lines between $\gamma(x - \epsilon)$ and $\gamma(x + \epsilon)$. The pair of the corresponding functions are denoted by (u_ϵ, w_ϵ).

One can regard that the pairs of the functions from (u_ϵ, w_ϵ) to (u, w) represents 'time evolution' of curves with discrete time variable. Below we verify that there are the expansions of these functions with respect to ϵ such as $u_\epsilon = u + \epsilon^2 F(v_x, u_{xx})$ and $v_\epsilon = v + \epsilon^2 G(v_{xx}, u, u_x, u_{xxx})$, where both F and G are polynomials, and u_{xx} implies differentiation twice by x. In such cases we obtain the system of the equations as a continuous limit:

$$
\begin{cases}
\dot{u} = F(v_x, u_{xx}), \\
\dot{v} = G(v_{xx}, u, u_x, u_{xxx}).
\end{cases}
$$

Theorem 5.6. *The continuous limit of the pentagram map T induces the following equation:*

$$u_{tt} = \frac{(u^2)_{xx}}{6} - \frac{u_{4x}}{12}.$$

Proof. Step 1: Let Γ_ϵ be the lift of γ_ϵ. It satisfies two relations:

$$\begin{cases} \det(\Gamma(x+\epsilon), \Gamma(x-\epsilon), \Gamma_\epsilon(x)) = 0, \\ \det(\Gamma_\epsilon(x), \Gamma(x+\epsilon) - \Gamma(x-\epsilon), \Gamma'_\epsilon(x)) = 0. \end{cases}$$

Let us expand $\Gamma_\epsilon = \Gamma + \epsilon A + \epsilon^2 B + (\epsilon^3)$, and insert into the above relations. Then

$$\begin{cases} \det(\Gamma + \epsilon\Gamma', \Gamma - \epsilon\Gamma', \Gamma + \epsilon A) = 2\epsilon^2 \det(\Gamma', \Gamma, A) = 0, \\ \det(\Gamma + \epsilon A, 2\epsilon\Gamma', \Gamma' + \epsilon A') = 2\epsilon^2 \det(\Gamma, \Gamma', A') + 2\epsilon^3 \det(A, \Gamma', A') = 0. \end{cases}$$

Let us denote $A = f\Gamma + g\Gamma' + h\Gamma''$ for some functions f, g, h. Then $h = 0$ holds from the first equality. From the second, $g = 0$ also holds. So $A = f\Gamma$.

By the same way, it follows that $B = \frac{1}{2}\Gamma'' + g\Gamma$. So

$$\Gamma_\epsilon = (1 + \epsilon f + \epsilon^2 g)\Gamma + \frac{\epsilon^2}{2}\Gamma'' + (\epsilon^3).$$

Step 2: We claim that $f \equiv 0$ and $g(x) = \frac{u(x)}{3}$ hold. In fact let us use the condition:

$$\det(\Gamma_\epsilon(x), \Gamma'_\epsilon(x), \Gamma''_\epsilon(x)) = 1$$

and insert the above expression. Then we have:

$$1 = \det\left(\alpha\Gamma + \frac{\epsilon^2}{2}\Gamma'', \alpha'\Gamma + \alpha\Gamma' + \frac{\epsilon^2}{2}\Gamma''', \alpha''\Gamma + 2\alpha'\Gamma' + \alpha\Gamma'' + \frac{\epsilon^2}{2}\Gamma''''\right)$$

where $\alpha = 1 + \epsilon f + \epsilon^2 g$.

Then we compare the coefficients of ϵ. $f = 0$ follows from the first order on ϵ, and the equality $g(x) = \frac{u(x)}{3}$ follows from the second order coefficient.

Step 3: Now we have the expression:

$$\Gamma'''_\epsilon(x) + u_\epsilon(x)\Gamma'_\epsilon(x) + v_\epsilon(x)\Gamma_\epsilon = 0.$$

Then we compute its derivatives:

$$\Gamma_\epsilon = \alpha\Gamma + \frac{\epsilon^2}{2}\Gamma'',$$

$$\Gamma'_\epsilon = \epsilon^2\left(\frac{u'}{3} - \frac{v}{2}\right)\Gamma + \left(1 - \epsilon^2\frac{u}{6}\right)\Gamma',$$

$$\Gamma'''_\epsilon = \left\{-v\left(1 - \frac{\epsilon^2 u}{6}\right) + \epsilon^2\left(\frac{u'''}{3} - \frac{v''}{2}\right)\right\}\Gamma$$

$$+ \left\{-u\left(1 - \frac{\epsilon^2 u}{6}\right) + \epsilon^2\left(\frac{u''}{2} - v'\right)\right\}\Gamma' + \dots$$

By comparisons of the coefficients of Γ', we obtain:

$$u_\epsilon = u + \epsilon^2\left(v' - \frac{u''}{2}\right).$$

Next by comparison of the coefficients of Γ, we obtain:

$$v_\epsilon = v + \epsilon^2\left(\frac{v''}{2} - \frac{uu'}{3} - \frac{u'''}{3}\right).$$

As a continuous limit, we obtain the system of the equations:

$$\begin{cases} u_t = v_x - \frac{u_{xx}}{2} \\ v_t = \frac{v''}{2} - \frac{uu'}{3} - \frac{u'''}{3}. \end{cases}$$

Let us put $u_t = (v - \frac{u_x}{2})_x \equiv w_x$, and compute:

$$w_{tx} = \left(v_t - \frac{u_{xt}}{2}\right)_x = \frac{v_{3x}}{2} - \frac{(uu_x)_x}{3} - \frac{u_{4x}}{3} - \frac{v_{3x}}{2} + \frac{u_{4x}}{4}$$

$$= \frac{(u^2)_{xx}}{6} - \frac{u_{4x}}{12}.$$

So we have induced into a single equation from the above system:

$$u_{tt} = \frac{(u^2)_{xx}}{6} - \frac{u_{4x}}{12}.$$

$\square$

Remark 5.5. Let us rewrite the equation as:

$$A = \left(\frac{d}{dx}\right)^3 + \frac{1}{2}\left(u(x)\frac{d}{dx} + \frac{d}{dx}u(x)\right) + w(x)$$

where $w(x) = v(x) - \frac{u'(x)}{2}$.

The pair (u, v) can be understood as a continuous analog of the coordinate (a_i, b_i) in 5.4.

References

The theory of the pentagram map was initiated by Schwartz [Sch1], [Sch2], [Sch3]. For basic integrable systems, refer to [KP] which also includes classical background. See [OST1], [OST2], [Sol] for integrability of the pentagram map. See [Gri] for the relation with cluster algebras, and [Kat10] for the pentagram automaton and application of the comparison method. The relation between the pentagram map and the Boussinesq equation is found in [OST1].

PART 2

Evolutional dynamics

Chapter 6

State dynamics

Roughly speaking an automaton is a rule defined over finite sets. Even though it is so abstract, it can produce quite interesting objects in the fields of not only algebra but also geometry and analysis. In chapter 7, we shall treat automata groups in relation to geometric group theory, and cell automata of integrable systems in relation to mathematical physics.

The most general form of such abstract rules will be the Turing machine, but in this book we will introduce a more restrictive class of automata which is enough to apply to such objects passing through tropical geometry.

6.1 Dynamics with states

Let us introduce a class of dynamical systems. Let Q be a set called *states* and X be a space. X will mostly be a set of infinite sequences over some sets.

Definition 6.1. A *state dynamics* on X is the dynamical system such that every $q \in Q$ determines an action on X. We call q the initial state.

A *deterministic automaton* consists of a finite rule given by functions over finite sets. It can create quite complicated state dynamics on the set of sequences of alphabets.

Let S be another set called *alphabets*, and consider all the sets of infinite sequences:

$$X_S = \{(s_0, s_1, \dots) : s_i \in S\}.$$

Let $S^m = S \times \cdots \times S$ be the product set of S by m times, and consider an automaton $\mathbf{A}$ which is given by a pair of functions:

$$\left\{ \begin{array}{l} \psi : Q \times S^{\alpha+1} \to S, \\ \phi : Q \times S^{\beta+1} \to Q \end{array} \right.$$

where $\alpha, \beta \geq 0$.

Definition 6.2. The state dynamics by (ψ, ϕ):

$$\mathbf{A}_q : X_S \to X_S, \quad \mathbf{A}_q(\bar{s}) = (s_0', s_1', \dots)$$

is defined inductively by:

$$\left\{ \begin{array}{l} s_i' = \psi(q_i, s_i, \dots, s_{i+\alpha}), \\ q_{i+1} = \phi(q_i, s_i, \dots, s_{i+\beta}) \quad (q_0 = q) \end{array} \right.$$

for any $q \in Q$ and $\bar{s} = (s_0, s_1, \dots) \in X_S$.

We call $\mathbf{A}_q$ the initial automaton, if we specify the initial state $q_0 = q \in Q$.

Remark 6.1. Besides the dynamics over X_S, the hidden dynamics by the state sets:

$$q_0 \to q_1 \to q_2 \to \cdots$$

play a quite important role.

Any sequence $\bar{q}^j = (q^0, \dots, q^j) \in X_Q^{j+1}$ of the initial states give dynamics by compositions:

$$\mathbf{A}_{\bar{q}^j} = \mathbf{A}_{q^j} \circ \cdots \circ \mathbf{A}_{q^0} : X_S \to X_S.$$

Definition 6.3. Two automata $\mathbf{A}$ and $\mathbf{A}'$ are said to be equivalent over the same alphabets A and the states Q, if they give the same dynamics:

$$\mathbf{A}_q = \mathbf{A}_q' : X_S \to X_S$$

for all $q \in Q$.

Two different automata can give equivalent state dynamics. In such case, the dynamics $\mathbf{A}_q$ are the same, but dynamics of the states behind can be very different.

6.2   Examples

The above class of state dynamics contains two important cases which appear in geometry and analysis. In chapters 7 and 10, we shall explain more detailed aspects of these subjects, but here we will just give some examples.

6.2.1 Cell automata of integrable systems

Many integrable systems are described by evolutional PDE systems, and there have been various studies to discretize such dynamical systems. One of the typical instance is the *Lotka–Volterra cell automaton*, whose dynamics is given by the rule:

$$s_i' = \varphi(s_{i-1}', s_i, s_{i+1}) = s_i + \max(0, s_{i+1}) - \max(0, s_{i-1}').$$

Let us rewrite this using state dynamics. For $S = Q$ with the initial state $q_0 = q$, consider the state dynamics with $\psi : Q \times S \to S, \phi : Q \times S^2 \to Q$ by:

$$\begin{cases} \psi(q, s) = q, \\ \phi(q, s_0, s_1) = s_0 + \max(0, s_1) - \max(0, q). \end{cases}$$

One can assign $s_i' = q_i$ with $s_0' = q$, which describes the above automaton.

There are other cases of the integrable systems given by cell automata. We will describe some of important cases of the integrable systems in chapter 10.

6.2.2 Automata group

As a special case $\alpha = \beta = 0$, the state dynamics of the form:

$$\begin{cases} \psi : Q \times S \to S, \\ \phi : Q \times S \to Q \end{cases}$$

is called *Mealy automaton*.

If we identify X_S with the rooted regular tree, then the state dynamics $\mathbf{A}_q$ give the semi-group actions on the tree, since the actions can be restricted level-setwisely.

They are invertible actions if and only if:

$$\psi : (q, \ ) : S \cong S$$

are isomorphic as sets for all $q \in Q$. In the invertible case, the group generated by these states is called the *automata group* given by the automaton (ψ, ϕ).

Automata groups solve many important problems in group theory. Later we will treat this particular class of Mealy automata in detail.

6.2.3 *Automata twisted group*

If $\psi : (q, \quad) : S \cong S$ is not one to one onto for some $q \in Q$, then the action $\mathbf{A}_q : X_S \to X_S$ is not invertible. In chapter 10, we will treat the integrable cell automata called box-ball systems (BBS) with carriers in detail. A box-ball system consists of an important class of integrable Mealy automata. They preserve the number of balls and hence conserve the system. On the other hand they do not satisfy the invertibility, and hence they can give rise to semi-group actions on the tree. In fact it seems quite rare to hold these two properties at the same time.

BBS systems still satisfy the property so that the pairs:

$$(\psi, \phi) : S \times Q \cong S \times Q$$

give rise to isomorphisms. We call such a property *twisted invertibility*.

Remark 6.2. Twisted invertibility on a Mealy automaton induces a quite important property which is stochasticity on all elements of approximation of the Markov operator (see proposition 11.2). Actually it gives a countable state Markov shift over the boundary of the tree on which the semi-group acts.

Recently S. Tsujimoto started a project to classify Mealy automata with small numbers of the states and alphabets by numerical computation. It gave a result that the number of the integrable Mealy automata with this property is surprisingly small. In 7.4, we will give a survey of this classification approach.

6.2.4 *Non Mealy automata group*

General state dynamics will not give actions on the trees, but still it does give actions on the boundary of the tree which is identified with X_S.

Here we give an example of a non Mealy automaton which gives the isomorphic actions on the boundary of the tree, but which are not induced from the actions on the tree itself.

One of the most basic properties commonly shared among all Mealy automata is that the groups are residually finite, since they can induce the restrictions level set wisely on the tree. In our case, the groups do not induce such restriction.

Let us consider an automaton $\mathbf{A}$ given by the two functions:

$$\left\{ \begin{array}{l} \psi : Q \times S^{\alpha+1} \to S, \\ \phi : Q \times S^{\beta+1} \to Q \end{array} \right.$$

and consider the corresponding state dynamics $\mathbf{A}_q : X_S \to X_S$ for each $q \in Q$. $\mathbf{A}_q : \partial T_m \to \partial T_m$ does not induce the action on the tree T_m.

Let us say that $\mathbf{A}$ is *invertible*, if $\mathbf{A}_q : X_S \cong X_S$ are isomorphisms for all $q \in Q$. An invertibe automaton $\mathbf{A}$ gives automorphisms $\mathbf{A}_q : \partial T_m \cong \partial T_m$, and the group generated by the set of states is also denoted by $G(\mathbf{A})$:

$$G(\mathbf{A}) = \text{gen}\ \{\mathbf{A}_q : \partial T_m \cong \partial T_m : q \in Q\}$$

which is a subgroup of the boundary automorphism group $\text{Aut}(\partial T_m)$.

Example 6.1. The following automaton is not Mealy, since dynamics are not determined level set wisely.

Let us choose $\alpha = \beta = 1$. Let $S = \{s_0, s_1\}$, $Q = \{q^0, q^1\}$ and ϵ_0 be the permutation between two elements. Let us consider the functions:

$$\psi(q^1, \ , s) = \epsilon_0, \quad \psi(q^0, \ , s) = \begin{cases} \epsilon_0 & s = s_0 \\ id & s = s_1 \end{cases}$$

$$\phi(\ , s, s') = \epsilon_0$$

where $s, s' \in S$.

Lemma 6.1. *The state dynamics by this automaton is invertible.*

Proof. Let us consider $\psi(q^0, \ , \ )$, and suppose $(q^0, s, t) \neq (q^0, s', t')$. The equalities $\psi(q^0, s, t) = \psi(q^0, s', t')$ hold only when $(s, s') = (s_1, s_0)$ or (s_0, s_1) and $t \neq t'$.

On the other hand at the next level, $\phi(q^0, s, t) = \phi(q^0, s', t') = q^1$ holds, and so:

$$\psi(q^1, t, \ ) \neq \psi(q^1, t', \ ).$$

This implies $\mathbf{A}_{q^0} : \partial T_m \to \partial T_m$ is injective.

Since $\psi(q^1, \ , \ )$ does not depend on the second variable, injectivity of $\mathbf{A}_{q^1} : \partial T_m \to \partial T_m$ follows from the one of $\mathbf{A}_{q^0}$.

Let us consider surjectivity, and take any $(x'_0, x'_1, \dots) \in X_S$. We seek for some elements $(x_0, x_1 \dots) \in X_S$ with $\mathbf{A}_q(x_0, x_1, \dots) = (x'_0, x'_1, \dots)$. Notice that the states change periodically as $(q^0, q^1, q^0, q^1, \dots)$ or $(q^1, q^0, q^1, \dots)$.

Both cases can be considered similarly, and we treat the first case only. Firstly we choose $x_{2i+1} = x'_{2i+1} + 1 \mod 2$ for all $i \geq 0$. Then we can choose x_{2i} uniquely so that the equalities $\psi(q^0, x_{2i}, x_{2i+1}) = x'_{2i}$ hold for all i. Thus $\mathbf{A}_q$ are surjective, and so they are isomorphisms. $\square$

6.3   State dynamics

Let:

$$\begin{cases} \varphi : \mathbb{R}^{\alpha+2} \to \mathbb{R}, \\ \psi : \mathbb{R}^{\beta+2} \to \mathbb{R} \end{cases}$$

be two piecewise-linear functions which admit their presentations by relative (max, +)-functions:

$$\psi(\bar{r}) = \max(\alpha_1 + \bar{a}_1\bar{r}, \ldots, \alpha_{\alpha+1} + \bar{a}_{\alpha+2}\bar{r}) - \max(\beta_1 + \bar{b}_1\bar{r}, \ldots, \beta_{\alpha+1} + \bar{b}_{\alpha+2}\bar{r}),$$

$$\phi(\bar{l}) = \max(\gamma_1 + \bar{c}_1\bar{l}, \ldots, \gamma_{\beta+1} + \bar{c}_{\beta+2}\bar{l}) - \max(\delta_1 + \bar{d}_1\bar{l}, \ldots, \delta_{\beta+1} + \bar{d}_{\beta+2}\bar{l})$$

where $\alpha, \beta, \gamma, \delta \in \mathbb{R}$, $\bar{a}, \bar{b} \in \mathbb{R}^{\alpha+1}$ and $\bar{c}, \bar{d} \in \mathbb{R}^{\beta+1}$ are all constants.

Later on we denote $M = \max(M_\psi, M_\phi)$ and $c = \max(c_\psi, c_\phi)$ as the larger ones of the numbers of the components and the Lipschitz constants for ψ and ϕ respectively.

Let us take initial sequences in $\mathbb{R}$:

$$\{x_i\}_{i \geq 0}, \qquad \{y^j\}_{j \geq 0}.$$

Definition 6.4. The state dynamics with respect to the pair (ψ, ϕ) is given by the discrete dynamics inductively defined by the formulas:

$$\begin{cases} x_i^{j+1} = \psi(y_i^j, x_i^j, \ldots, x_{i+\alpha}^j), \\ y_{i+1}^j = \phi(y_i^j, x_i^j, \ldots, x_{i+\beta}^j) \end{cases}$$

where the initial values are given by $x_i^0 = x_i$ and $y_0^j = y^j$ for $i, j = 0, 1, 2, \ldots$

Firstly one determines $y_1^0, y_2^0, y_3^0, \ldots$ by the use of the initial data. Then determine $x_0^1, x_1^1, x_2^1, \ldots$ by the use of $\{y_i^0\}_i$ and the initial data. Next determine $y_1^1, y_2^1, y_3^1, \ldots$, and $x_0^2, x_1^2, x_2^2, \ldots$, and so on.

The dynamical distribution is of the form:

$$\begin{array}{llll} \mathbf{x_0^0} & \mathbf{x_1^0} & \mathbf{x_2^0} & \mathbf{x_3^0} \ \cdots \\ x_0^1 & x_1^1 & x_2^1 & x_3^1 \ \cdots \\ x_0^2 & x_1^2 & x_2^2 & x_3^2 \ \cdots \end{array}$$

while we have the hidden dynamics by states:

$$\begin{array}{llll} \mathbf{y_0^0} & y_1^0 & y_2^0 & y_3^0 \ \cdots \\ \mathbf{y_0^1} & y_1^1 & y_2^1 & y_3^1 \ \cdots \\ \mathbf{y_0^2} & y_1^2 & y_2^2 & y_3^2 \ \cdots \end{array}$$

The bold characters are the initial data.

Recall the dynamics of automata in 6.2, say for the automata group case. The row of x variables from $\{x_i^0\}_i$ to $\{x_i^1\}_i$ corresponds to the action by $\mathbf{A}_{y^0}$. Hence $\{x_i^0\}_i$ to $\{x_i^l\}_i$ corresponds to the action by $\mathbf{A}_{(y^0 y^1 \ldots y^{l-1})} = \mathbf{A}_{y^{l-1}} \circ \cdots \circ \mathbf{A}_{y^0}$.

Let (ψ_t, ϕ_t) be the tropical correspondences of the pair (ψ, ϕ). One can also consider another state dynamics with the same rule using (ψ_t, ϕ_t) instead of the pair. Recall we have obtained the uniform comparison estimates on the orbits of tropical correspondences in chapter 3. In chapter 6, we induce the corresponding uniform comparison estimates for the state dynamics.

6.4 Preliminary estimates on the orbits of the state dynamics

Let us induce some preliminary estimates for later purposes.

Firstly let us start analyzing the orbits $\{y_i^0\}_j$, since their treatment is relatively simple compared with $\{x_i^j\}_{i,j}$.

Let $\varphi : \mathbb{R}^n \to \mathbb{R}$ be a relative (max, +)-function of n variables, with the number of the components M and the Lipschitz constant c. Let φ_t be the corresponding function.

Now we consider the dynamics of the states, and choose any initial data $(x_0, x_1, \ldots) \in \mathbb{R}^{\mathbb{N}}$ and $y_0 \in \mathbb{R}$. Then one considers two discrete dynamics defined inductively by the iterations for $i \geq 0$:

$$y_{i+1} = \varphi(y_i, x_i, \ldots, x_{i+\beta}),$$
$$y'_{i+1} = \varphi_t(y'_i, x_i, \ldots, x_{i+\beta})$$

where $y'_0 = y_0$. We denote $\bar{x}_i = (x_i, \ldots, x_{i+\beta})$ for simplicity of the notation.

Recall the polynomials of degree i in 3.3 with the equality:

$$P_i(c) = \frac{c^{i+1} - 1}{c - 1}, \qquad cP_i(c) + 1 = P_{i+1}(c).$$

Lemma 6.2. *The uniform estimates:*

$$|y_i - y'_i| \leq P_{i-1}(c) \log_t M$$

hold for all $i \geq 1$.

Proof. Firstly the estimates $|y'_1 - y_1| = |\varphi_t(y_0, \bar{x}_0) - \varphi(y_0, \bar{x}_0)| \leq \log_t M$ hold by lemma 3.4.

Next we have the estimates:

$$
\begin{aligned}
|y_2 - y_2'| &= |\varphi(y_1, \bar{x}_1) - \varphi_t(y_1', \bar{x}_1)| \\
&\leq |\varphi(y_1, \bar{x}_1) - \varphi(y_1', \bar{x}_1)| + |\varphi(y_1', \bar{x}_1) - \varphi_t(y_1', \bar{x}_1)| \\
&\leq c|y_1 - y_1'| + \log_t M \leq (c+1)\log_t M.
\end{aligned}
$$

Similarly we have the following estimates:

$$
\begin{aligned}
|y_3 - y_3'| &= |\varphi(y_2, \bar{x}_2) - \varphi_t(y_2', \bar{x}_2)| \\
&\leq |\varphi(y_2, \bar{x}_2) - \varphi(y_2', \bar{x}_2)| + |\varphi(y_2', \bar{x}_2) - \varphi_t(y_2', \bar{x}_2)| \\
&\leq [c(c+1) + 1]\log_t M.
\end{aligned}
$$

By iterating the same estimates, one obtains the conclusion.

□

The estimates for the dynamics of $\{x_i^j\}_{i,j}$ involve more complicated analysis. As preliminaries, we verify some general estimates which will be used later.

For four sequences $\{p_i\}_i, \{q^j\}_j, \{x_i\}_i, \{y^j\}_j$ let us introduce the numbers:

$$
|\{p_i, x_i\}_i; \{q^j, y^j\}_j| \equiv \sup_{i,j}\{|p_i - x_i|, \ |y^j - q^j|\}.
$$

For the Lipschitz constants c, let us put:

$$
\tilde{c} = \max(c, 1).
$$

The following type of estimates is applied when we consider Mealy automata. The general cases are treated after this.

Lemma 6.3. *Let us consider four sequences:*

$$
\{p_i^j\}_{i,j \geq 0}, \ \{q_i^j\}_{i,j \geq 0}, \ \{x_i^j\}_{i,j \geq 0}, \ \{y_i^j\}_{i,j \geq 0}.
$$

Suppose these sequences satisfy the following estimates:

$$
|p_i^{j+1} - x_i^{j+1}|, \ |q_{i+1}^j - y_{i+1}^j| \leq c \max(|q_i^j - y_i^j|, |p_i^j - x_i^j|) + T
$$

for some $T \in \mathbb{R}$ and $c \geq 0$.

Then they satisfy the estimates for all $i, j \geq 0$:

$$
|p_i^{j+1} - x_i^{j+1}|, \ |q_{i+1}^j - y_{i+1}^j| \leq P_{i+j}(c)T + \tilde{c}^{i+j+1}|\{p_i^0, x_i^0\}_i; \{q_0^j, y_0^j\}_j|.
$$

In particular:

(1) If they satisfy the same initial conditions:

$$p_i^0 = x_i^0, \quad q_0^j = y_0^j \qquad (i, j \geq 0)$$

then the uniform estimates hold:

$$|p_i^{j+1} - x_i^{j+1}|, \quad |q_{i+1}^j - y_{i+1}^j| \leq P_{i+j}(c)T.$$

(2) $|p_i^{j+1} - x_i^{j+1}|$ and $|q_{i+1}^j - y_{i+1}^j|$ are both uniformly bounded, if $c < 1$ holds and $|\{p_i^0, x_i^0\}_i; \{q_0^j, y_0^j\}_j|$ is finite.

Proof. We verify the conclusion by induction on $i + j \geq 1$.

For $i + j = 0$, the estimates:

$$|p_0^1 - x_0^1|, \quad |q_1^0 - y_1^0| \leq c \max(|q_0^0 - y_0^0|, \ |p_0^0 - x_0^0|) + T$$
$$\leq P_1(c)T + \bar{c}|\{p_i^0, x_i^0\}_i; \{q_0^j, y_0^j\}_j|$$

hold. So the conclusion follows for $i + j = 1$.

Suppose the conclusion holds for all (i, j) with $i + j \leq N$, and take any (i, j) with $i + j = N + 1$.

Firstly assume both $i, j \neq 0$. Then $1 \leq i, j \leq N$ and hence the estimates hold:

$$|p_i^{j+1} - x_i^{j+1}|, \quad |q_{i+1}^j - y_{i+1}^j| \leq c \max(|q_i^j - y_i^j|, \ |p_i^j - x_i^j|) + T$$
$$\leq c(P_{i+j-1}(c)T + \tilde{c}^{i+j}|\{p_i^0, x_i^0\}_i; \{q_0^j, y_0^j\}_j|) + T$$
$$\leq P_{i+j}(c)T + \tilde{c}^{i+j+1}|\{p_i^0, x_i^0\}_i; \{q_0^j, y_0^j\}_j|.$$

Next suppose $j = 0$ and $i = N + 1$. Then:

$$|p_i^1 - x_i^1|, \quad |q_{i+1}^0 - y_{i+1}^0| \leq c \max(|q_i^0 - y_i^0|, \ |p_i^0 - x_i^0|) + T$$
$$\leq c \max(P_{i-1}(c)T + \tilde{c}^i|\{p_i^0, x_i^0\}_i; \{q_0^j, y_0^j\}_j|, |\{p_i^0, x_i^0\}_i; \{q_0^j, y_0^j\}_j|) + T$$
$$\leq P_{i+j}(c)T + \tilde{c}^{i+j+1}|\{p_i^0, x_i^0\}_i; \{q_0^j, y_0^j\}_j|$$

since the estimates hold:

$$|p_i^0 - x_i^0| \leq |\{p_i^0, x_i^0\}_i; \{q_0^j, y_0^j\}_j| \leq \tilde{c}^i|\{p_i^0, x_i^0\}_i; \{q_0^j, y_0^j\}_j|.$$

We can treat the case $i = 0$ by the same way. Thus we have verified the claim for $i + j \leq N + 1$. This finishes the induction step.

$\square$

Let us take four sequences $\{p_i\}_i, \{q^j\}_j, \{x_i\}_i, \{y^j\}_j$ as before. Let $\alpha, \beta \geq 0$ be constants and put:

$$\gamma = \max(\alpha, \beta).$$

Now we verify the general cases which can be applied to estimate the orbits of the state dynamics. The argument is more complicated than the above lemma.

Proposition 6.1. *Suppose these satisfy the following estimates:*

$$|p_i^{j+1} - x_i^{j+1}| \leq$$
$$c \max(|q_i^j - y_i^j|, |p_i^j - x_i^j|, |p_{i+1}^j - x_{i+1}^j|, \ldots, |p_{i+\alpha}^j - x_{i+\alpha}^j|) + T,$$
$$|q_{i+1}^j - y_{i+1}^j| \leq$$
$$c \max(|q_i^j - y_i^j|, |p_i^j - x_i^j|, |p_{i+1}^j - x_{i+1}^j|, \ldots, |p_{i+\beta}^j - x_{i+\beta}^j|) + T.$$

Then they satisfy the estimates:

$$|p_i^{j+1} - x_i^{j+1}|, \ |q_{i+1}^j - y_{i+1}^j|$$
$$\leq P_{i+j(\gamma+1)}(c)T + \tilde{c}^{i+1+j(\gamma+1)}|\{p_i^0, x_i^0\}_i; \{q_0^j, y_0^j\}_j|.$$

Proof. We split the proof into several steps.

Step 1: Firstly we claim that the estimates below hold by induction on i:

$$|q_i^0 - y_i^0| \leq P_{i-1}(c)T + \tilde{c}^i|\{p_i^0, x_i^0\}_i; \{q_0^j, y_0^j\}_j|.$$

For $i = 1$, the estimate $|q_1^0 - y_1^0| \leq T + \bar{c}|\{p_i^0, x_i^0\}_i; \{q_0^j, y_0^j\}_j|$ holds by assumption.

Suppose the claim holds up to i. Then:

$$|q_{i+1}^0 - y_{i+1}^0| \leq c \max(|q_i^0 - y_i^0|, |p_i^0 - x_i^0|, \ldots, |p_{i+\beta}^0 - y_{i+\beta}^0|) + T$$
$$\leq c(P_{i-1}(c)T + \tilde{c}^i|\{p_i^0, x_i^0\}_i; \{q_0^j, y_0^j\}_j|) + T$$
$$\leq P_i(c)T + \tilde{c}^{i+1}|\{p_i^0, x_i^0\}_i; \{q_0^j, y_0^j\}_j|.$$

Thus it holds up to $i + 1$. This verifies the claim.

Then we have the estimates:

$$|p_i^1 - x_i^1| \leq c \max(|q_i^0 - y_i^0|, |p_i^0 - x_i^0|, \ldots, |p_{i+\alpha}^0 - y_{i+\alpha}^0|) + T$$
$$\leq c(P_{i-1}(c)T + \tilde{c}^i|\{p_i^0, x_i^0\}_i; \{q_0^j, y_0^j\}_j|) + T$$
$$\leq P_i(c)T + \tilde{c}^{i+1}|\{p_i^0, x_i^0\}_i; \{q_0^j, y_0^j\}_j|.$$

Step 2: Next we claim that the estimates hold:

$$|q_i^1 - y_i^1| \leq P_{i+\beta}(c)T + \tilde{c}^{i+\beta+1}|\{p_i^0, x_i^0\}_i; \{q_0^j, y_0^j\}_j|.$$

We proceed by induction on i. For $i = 0$, the estimates:

$$|q_0^1 - y_0^1| \leq |\{p_i^0, x_i^0\}_i; \{q_0^j, y_0^j\}_j| \leq P_\beta(c)T + \tilde{c}^{\beta+1}|\{p_i^0, x_i^0\}_i; \{q_0^j, y_0^j\}_j|$$

hold by definition, since $\tilde{c} \geq 1$ holds.

Suppose the above estimates hold up to $i \geq 0$. Then:

$$\begin{aligned}
|q_{i+1}^1 - y_{i+1}^1| &\leq c \max(|q_i^1 - y_i^1|, |p_i^1 - x_i^1|, \ldots, |p_{i+\beta}^1 - x_{i+\beta}^1|) + T \\
&\leq c \max(|q_i^1 - y_i^1|, P_{i+\beta}(c)T + \tilde{c}^{i+\beta+1}|\{p_i^0, x_i^0\}_i; \{q_0^j, y_0^j\}_j|) + T \\
&\leq c(P_{i+\beta}(c)T + \tilde{c}^{i+\beta+1}|\{p_i^0, x_i^0\}_i; \{q_0^j, y_0^j\}_j|) + T \\
&= P_{i+\beta+1}(c)T + \tilde{c}^{i+\beta+2}|\{p_i^0, x_i^0\}_i; \{q_0^j, y_0^j\}_j|
\end{aligned}$$

where we used step 1 at the second inequalities. So the above estimates also hold for $i+1$, and we have verified the claim.

Then we have the estimates:

$$\begin{aligned}
|p_i^2 - x_i^2| &\leq c \max(|q_i^1 - y_i^1|, |p_i^1 - x_i^1|, .., |p_{i+\alpha}^1 - x_{i+\alpha}^1|) + T \\
&\leq c \max(P_{i+\beta}(c)T + \tilde{c}^{i+\beta+1}|\{p_i^0, x_i^0\}_i; \{q_0^j, y_0^j\}_j|, \\
&\qquad P_{i+\alpha}(c)T + \tilde{c}^{i+\alpha+1}|\{p_i^0, x_i^0\}_i; \{q_0^j, y_0^j\}_j|) + T \\
&= P_{i+\gamma+1}(c)T + \tilde{c}^{i+\gamma+2}|\{p_i^0, x_i^0\}_i; \{q_0^j, y_0^j\}_j|.
\end{aligned}$$

Step 3: Let us verify the estimates for the general case by induction on j. In steps $1, 2$, we have verified the conclusions for $j \leq 1$ and all $i \geq 0$.

Suppose the conclusions hold up to $j - 1 \geq 1$ and all $i \geq 0$. Firstly let us consider the pair (y_i^j, q_i^j). We claim that the estimates hold for all $i \geq 0$:

$$|q_i^j - y_i^j| \leq P_{i-1+j(\gamma+1)}(c)T + \tilde{c}^{i+j(\gamma+1)}|\{p_i^0, x_i^0\}_i; \{q_0^j, y_0^j\}_j|.$$

The estimates $|q_0^j - y_0^j| \leq |\{p_i^0, x_i^0\}_i; \{q_0^j, y_0^j\}_j|$ hold by definition. Let us proceed by induction on i. Suppose the above estimates hold up to $i \geq 0$. Then:

$$\begin{aligned}
|q_{i+1}^j - y_{i+1}^j| &\leq c \max(|q_i^j - y_i^j|, |p_i^j - x_i^j|, \ldots, |p_{i+\beta}^j - x_{i+\beta}^j|) + T \\
&\leq c \max(P_{i-1+j(\gamma+1)}(c)T + \tilde{c}^{i+j(\gamma+1)}|\{p_i^0, x_i^0\}_i; \{q_0^j, y_0^j\}_j|, \\
&\qquad P_{i+\beta+(j-1)(\gamma+1)}(c)T + \tilde{c}^{i+\beta+1+(j-1)(\gamma+1)}|\{p_i^0, x_i^0\}_i; \{q_0^j, y_0^j\}_j|) + T \\
&= P_{i+j(\gamma+1)}(c)T + \tilde{c}^{i+1+j(\gamma+1)}|\{p_i^0, x_i^0\}_i; \{q_0^j, y_0^j\}_j|.
\end{aligned}$$

So the estimates also hold for $i+1$, and we have verified the claim.

Then finally we claim that the estimates:

$$|p_i^{j+1} - x_i^{j+1}| \leq P_{i+j(\gamma+1)}(c)T + \tilde{c}^{i+1+j(\gamma+1)}|\{p_i^0, x_i^0\}_i; \{q_0^j, y_0^j\}_j|$$

hold. This follows from the following:

$$
\begin{aligned}
|p_i^{j+1} - x_i^{j+1}| &\leq c \max(|q_i^j - y_i^j|, |p_i^j - x_i^j|, \ldots, |p_{i+\alpha}^j - x_{i+\alpha}^j|) + T \\
&\leq c \max(P_{i-1+j(\gamma+1)}(c)T + \tilde{c}^{i+j(\gamma+1)} |\{p_i^0, x_i^0\}_i; \{q_0^j, y_0^j\}_j|, \\
&\qquad\quad P_{i+\alpha+(j-1)(\gamma+1)}(c)T + \tilde{c}^{i+\alpha+1+(j-1)(\gamma+1)} |\{p_i^0, x_i^0\}_i; \{q_0^j, y_0^j\}_j|) + T \\
&= P_{i+j(\gamma+1)}(c)T + \tilde{c}^{i+1+j(\gamma+1)} |\{p_i^0, x_i^0\}_i; \{q_0^j, y_0^j\}_j|.
\end{aligned}
$$

Thus we have verified the claim under the induction hypothesis up to $j - 1$.

□

6.5 Uniform estimates on the orbits for state dynamics

Let ψ and ϕ be a pair of relative (max, +)-functions, and c and M be the maximums of the Lipschitz constants and the numbers of their components respectively. Let ψ_t and ϕ_t be the tropical correspondences to ψ and ϕ respectively.

Let us start from the Mealy case and so suppose both ψ and ϕ have two variables, and consider the systems of the equations:

$$
\begin{cases}
x(i, j+1) = \psi(y(i,j), x(i,j)), \\
y(i+1, j) = \phi(y(i,j), x(i,j)),
\end{cases}
$$
$$
\begin{cases}
x'(i, j+1) = \psi_t(y'(i,j), x'(i,j)), \\
y'(i+1, j) = \phi_t(y'(i,j), x'(i,j))
\end{cases}
$$

with the same initial values $x'(i, 0) = x(i, 0) = x_i$ and $y'(0, j) = y(0, j) = y^j$ for all $i, j \geq 0$.

Lemma 6.4. *The uniform estimates hold:*

$$
|x(i, j+1) - x'(i, j+1)|, \ |y(i+1, j) - y'(i+1, j)| \ \leq P_{i+j}(c) \log_t M.
$$

Proof. By lemma 3.4, both the estimates hold:

$$|x(i, j + 1) - x'(i, j + 1)| = |\psi(y(i, j), x(i, j)) - \psi_t(y'(i, j), x'(i, j))|$$
$$\leq |\psi(y(i, j), x(i, j)) - \psi(y'(i, j), x'(i, j))|$$
$$+ |\psi(y'(i, j), x'(i, j)) - \psi_t(y'(i, j), x'(i, j))|$$
$$\leq c \max(|y(i, j) - y'(i, j)|, |x(i, j) - x'(i, j)|) + \log_t M,$$

$$|y(i + 1, j) - y'(i + 1, j)| = |\phi(y(i, j), x(i, j)) - \phi_t(y'(i, j), x'(i, j))|$$
$$\leq |\phi(y(i, j), x(i, j)) - \phi(y'(i, j), x'(i, j))|$$
$$+ |\phi(y'(i, j), x'(i, j)) - \phi_t(y'(i, j), x'(i, j))|$$
$$\leq c \max(|y(i, j) - y'(i, j)|, |x(i, j) - x'(i, j)|) + \log_t M.$$

By applying lemma 6.3 for $p_i^j = x'(i, j)$, $x(i, j) = x_i^j$, $q_i^j = y'(i, j)$ and $y_i^j = y(i, j)$ with $T = \log_t M$, one obtains the desired result.

$\square$

Now let us consider the general case. Let ψ and ϕ be a pair of relative (max, +)-functions.

Let us choose the initial data $\{x_i\}_{j \geq 0}$ and $\{y^j\}_{j \geq 0}$ respectively, and consider two state dynamics given by the systems of the equations:

$$\begin{cases} x(i, j + 1) = \psi(y(i, j), x(i, j), \dots, x(i + \alpha, j)) \\ y(i + 1, j) = \phi(y(i, j), x(i, j), \dots, x(i + \beta, j)), \end{cases}$$

$$\begin{cases} x'(i, j + 1) = \psi_t(y'(i, j), x'(i, j), \dots, x'(i + \alpha, j)) \\ y'(i + 1, j) = \phi_t(y'(i, j), x'(i, j), \dots x'(i + \beta, j)), \end{cases}$$

with the same initial values $x(i, 0) = x'(i, 0) = x_i$ and $y(0, j) = y'(0, j) = y^j$.

Proposition 6.2. *The uniform estimates hold:*

$$|x(i, j + 1) - x'(i, j + 1)|, \ |y(i + 1, j) - y'(i + 1, j)| \leq P_{i + j(\gamma + 1)}(c) \log_t M.$$

Proof. The proof is parallel to lemma 6.4. By lemma 3.4, one has the estimates:

$$|x(i, j+1) - x'(i, j+1)|$$
$$= |\psi(y(i,j), x(i,j), \ldots, x(i+\alpha, j)) - \psi_t(y'(i,j), x'(i,j), \ldots, x'(i+\alpha, j))|$$
$$\leq |\psi(y(i,j), x(i,j), \ldots, x(i+\alpha, j)) - \psi(y'(i,j), x'(i,j), \ldots, x'(i+\alpha, j))|$$
$$+ |\psi(y'(i,j), x'(i,j), \ldots, x'(i+\alpha, j)) - \psi_t(y'(i,j), x'(i,j), \ldots, x'(i+\alpha, j))|$$
$$\leq c \max(|y(i,j) - y'(i,j)|, |x(i,j) - x'(i,j)|,$$
$$\ldots, |x(i+\alpha, j) - x'(i+\alpha, j)|) + \log_t M,$$

$$|y(i+1, j) - y'(i+1, j)|$$
$$= |\phi(y(i,j), x(i,j), \ldots, x(i+\beta, j)) - \phi_t(y'(i,j), x'(i,j), \ldots, x'(i+\beta, j))|$$
$$\leq |\phi(y(i,j), x(i,j), \ldots, x(i+\beta, j)) - \phi(y'(i,j), x'(i,j), \ldots, x'(i+\beta, j))|$$
$$+ |\phi(y'(i,j), x'(i,j), \ldots, x'(i+\beta, j)) - \phi_t(y'(i,j), x'(i,j), \ldots, x'(i+\beta, j))|$$
$$\leq c \max(|y(i,j) - y'(i,j)|, |x(i,j) - x'(i,j)|,$$
$$\ldots, |x(i+\beta, j) - x'(i+\beta, j)|) + \log_t M.$$

By applying proposition 6.1 for $p_i^j = x'(i,j)$, $x(i,j) = x_i^j$, $q_i^j = y'(i,j)$ and $y_i^j = y(i,j)$ with $T = \log_t M$, one obtains the result.

$\square$

6.6 Initial value dependence

Let ϕ, ψ and M, c be as in 6.5, and consider the state dynamics:

$$\begin{cases} x(i, j+1) = \psi(y(i,j), x(i,j), \ldots, x(i+\alpha, j)), \\ y(i+1, j) = \phi(y(i,j), x(i,j), \ldots, x(i+\beta, j)). \end{cases}$$

We consider how the initial values influence the long time behavior of the dynamics. For $l = 1, 2$, let $\{x_i(l)\}_{j \geq 0}$ and $\{y^j(l)\}_{j \geq 0}$ be two initial data, and denote the corresponding solutions by $\{(x_l(i,j), y_l(i,j))\}_{l=1,2}$ with $x_l(i, 0) = x_i(l)$ and $y_l(0, j) = y^j(l)$. Recall $|\{x_i(1), x_i(2)\}_i; \{y^j(1), y^j(2)\}_j|$ in 6.4.

Lemma 6.5. *The estimates hold:*

$$|x_1(i, j+1) - x_2(i, j+1)|, \; |y_1(i+1, j) - y_2(i+1, j)|$$
$$\leq \tilde{c}^{i+1+j(\gamma+1)} |\{x_i(1), x_i(2)\}_i; \{y^j(1), y^j(2)\}_j|.$$

Proof. Let us consider the estimates:

$$|x_1(i, j+1) - x_2(i, j+1)| =$$
$$|\psi(y_1(i, j), x_1(i, j), \ldots, x_1(i+\alpha, j)) - \psi(y_2(i, j), x_2(i, j), \ldots, x_2(i+\alpha, j))|$$
$$\leq c \max(|y_1(i, j) - y_2(i, j)|, \ldots, |x_1(i+\alpha, j) - x_2(i+\alpha, j)|)$$

$$|y_1(i+1, j) - y_2(i+1, j)| =$$
$$|\psi(y_1(i, j), x_1(i, j), \ldots, x_1(i+\beta, j)) - \psi(y_2(i, j), x_2(i, j), \ldots, x_2(i+\beta, j))|$$
$$\leq c \max(|y_1(i, j) - y_2(i, j)|, \ldots, |x_1(i+\beta, j) - x_2(i+\beta, j)|).$$

Then by applying proposition 6.1 for $p_i^j = x_1(i, j)$, $x_i^j = x_2(i, j)$ and $q_i^j = y_1(i, j)$, $y_i^j = y_2(i, j)$ with $T = 0$, one obtains the desired estimates.

$\square$

6.7 Rational dynamics

Let $\varphi : \mathbb{R}^{\alpha+2} \to \mathbb{R}$ and $\psi : \mathbb{R}^{\beta+2} \to \mathbb{R}$ be relative $(\max, +)$-functions with the constants M and c as before.

Passing through the scale transform in tropical geometry, one obtains two parametrized rational functions f_t and g_t with respect to φ and ψ of the form respectively:

$$f_t(\bar{z}) = \frac{t^{\alpha_1} \bar{z}^{\bar{a}_1} + \cdots + t^{\alpha_{\alpha+2}} \bar{z}^{\bar{a}_{\alpha+2}}}{t^{\beta_1} \bar{z}^{\bar{b}_1} + \cdots + t^{\beta_{\alpha+2}} \bar{z}^{\bar{b}_{\alpha+2}}}, \quad g_t(\bar{w}) = \frac{t^{\gamma_1} \bar{w}^{c_1} + \cdots + t^{\gamma_{\beta+2}} \bar{w}^{\bar{c}_{\beta+2}}}{t^{\delta_1} \bar{w}^{\bar{d}_1} + \cdots + t^{\delta_{\beta+2}} \bar{w}^{d_{\beta+2}}}.$$

Let us take initial values:

$$0 < z_i < \infty, \quad 0 < w^j < \infty$$

for $i, j = 0, 1, 2, \ldots$

Definition 6.5. The dynamical system given by:

$$\begin{cases} z_i^{j+1} = f_t(w_i^j, z_i^j, \ldots, z_{i+\alpha}^j), \\ w_{i+1}^j = g_t(w_i^j, z_i^j, \ldots, z_{i+\beta}^j) \end{cases}$$

is called the state system of the rational dynamics, where the initial values are given by $z_i^0 = z_i$ and $w_0^j = w^j$.

Let ψ_t and ϕ_t be the tropical correspondences to ψ and ϕ respectively. Let us consider the state dynamics:

$$\begin{cases} x'(i, j+1) = \psi_t(y'(i,j), x'(i,j), \ldots, x(i+\alpha, j)) \\ y'(i+1, j) = \phi_t(y'(i,j), x'(i,j), \ldots x(i+\beta, j)) \end{cases}$$

with the initial values $x'(i,0) = \log_t z_i$ and $y'(0,j) = \log_t w^j$.

Lemma 6.6. *The equalities hold:*

$$x'(i,j) = \log_t z_i^j, \quad y'(i,j) = \log_t w_i^j.$$

Proof. This follows from proposition 2.1. $\qquad\qquad\qquad\qquad\qquad\square$

6.8 Analysis on the equivalent dynamics for simple case

Compare the contents here with 3.4.

Let φ^1 and φ^2 be relative (max, +)-functions, which are mutually equivalent with the same Lipschitz constants c. Let M be the larger one of the numbers of the components.

Let g_t^1 and g_t^2 be the corresponding elementary rational functions which are mutually tropically equivalent. Let us take $w_0 \in (0, \infty)$ and an initial sequence $\{z_0, z_1, \ldots\} \in \mathbb{R}_{>0}^{\mathbb{N}}$ of positive numbers. Then one considers the rational dynamics defined inductively by the iterations:

$$w_{i+1}(l) = g_t^l(w_i(l), z_i, \ldots, z_{i+\beta}) \qquad (l = 1, 2)$$

with $w_0(l) = w_0$. Notice that these orbits take positive values.

Lemma 6.7. *The uniform estimates hold for all $i \geq 0$:*

$$\left(\frac{w_i(1)}{w_i(2)}\right)^{\pm 1} \equiv \left\{ \frac{w_i(1)}{w_i(2)}, \frac{w_i(2)}{w_i(1)} \right\} \leq M^{2P_{i-1}(c)}.$$

Proof. Let us put $x_i = \log_t z_i$ and $y_0 = \log_t w_0$, and consider the discrete dynamics defined inductively by the iterations:

$$y_{i+1}(l) = \varphi^l(y_i(l), x_i, \ldots, x_{i+\beta}),$$
$$y'_{i+1}(l) = \varphi_t^l(y'_i(l), x_i, \ldots, x_{i+\beta})$$

with $y_0(l) = y'_0(l) = y_0$. The equalities:

$$y_i(1) = y_i(2)$$

hold for all $i \geq 0$, since φ^1 and φ^2 are mutually equivalent.

By proposition 2.1, the equalities hold:

$$y_i'(l) = \log_t w_i(l) \qquad (l = 1, 2)$$

Then the following estimates follow from lemma 3.5:

$$\log_t \left\{ \max \left(\frac{w_i(1)}{w_i(2)}, \frac{w_i(2)}{w_i(1)} \right) \right\} = |\log_t w_i(1) - \log_t w_i(2)| = |y_i'(1) - y_i'(2)|$$

$$\leq |y_i'(1) - y_i(1)| + |y_i(1) - y_i(2)| + |y_i(2) - y_i'(2)|$$

$$\leq 2P_{i-1}(c) \log_t M = \log_t M^{2P_{i-1}(c)}$$

where $P_i(c) = \frac{c^{i+1}-1}{c-1}$. Since $\log_t$ is increasing, these estimates imply the desired one.

$\square$

6.9 Analysis on the equivalent dynamics

Here we induce the basic estimates which will be used to apply the theory of automata groups to tropical geometry in chapter 9.

Recall in 6.3 that the state dynamics with respect to the pair (ϕ, ψ) requires initial data $\{x_i\}_i$ and $\{y^j\}_j$. Correspondingly the rational dynamics with respect to the pair (f_t, g_t) also requires initial data $\{z_i\}_i$ and $\{w^j\}_j$. Suppose we have two pairs (f_t^1, g_t^1) and (f_t^2, g_t^2) with their initial data. Then we have the comparison analysis of their orbit behaviors in the following.

For $l = 1, 2$, let $\{w^j(l)\}_{j \geq 0}$ and $\{z_i(l)\}_{i \geq 0}$ be sequences of positive numbers.

Definition 6.6. The initial rate is given by the positive number:

$$[\{(z_i(l), w^j(l))\}_{l=1}^2] = \sup_{i,j} \max\left(\frac{z_i(1)}{z_i(2)}, \frac{z_i(2)}{z_i(1)}, \frac{w^j(1)}{w^j(2)}, \frac{w^j(2)}{w^j(1)} \right) \geq 1.$$

Let us put $\log_t z_i(l) = x_i(l)$ and $\log_t w^j(l) = y^j(l)$. Then the equality holds (see 5.4):

$$\log_t [\{(z_i(l), w^j(l))\}_{l=1}^2] = |\{x_i(1), x_i(2)\}_i; \{y^j(1), y^j(2)\}_j|.$$

Let (ψ^1, ϕ^1) and (ψ^2, ϕ^2) be pairs of $(\max, +)$-functions so that $\psi^1 \sim \psi^2$ and $\phi^1 \sim \phi^2$ are mutually equivalent (same as functions but their presentations may be different). We say that (ψ^1, ϕ^1) and (ψ^2, ϕ^2) are *pairwisely equivalent*.

Let c_φ and M_φ be the Lipschitz constant and the number of the components for φ respectively. Then we put:

$$c = \max(c_{\psi^1}, c_{\psi^2}, c_{\phi^1}, c_{\phi^2}), \quad M = \max(M_{\psi^1}, M_{\psi^2}, M_{\phi^1}, M_{\phi^2}).$$

Let (f_t^1, g_t^1) and (f_t^2, g_t^2) be the elementary rational functions with respect to (ψ^1, ϕ^1) and (ψ^2, ϕ^2).

Let $\{w^j(l)\}_{j \geq 0}$ and $\{z_i(l)\}_{i \geq 0}$ be the initial sequences by positive numbers, and denote the solutions by $(z_i^j(l), w_i^j(l))$ to the rational dynamics:

$$z_i^{j+1}(l) = f_t^l(w_i^j(l), z_i^j(l), \ldots, z_{i+\alpha}^j(l)),$$
$$w_{i+1}^j(l) = g_t^l(w_i^j(l), z_i^j(l), \ldots, z_{i+\beta}^j(l))$$

with the initial values $z_i^0(l) = z_i(l)$ and $w_0^j(l) = w^j(l)$ respectively.

Proposition 6.3. *The uniform estimates hold:*

$$\max\left(\frac{z_i^{j+1}(1)}{z_i^{j+1}(2)}, \frac{z_i^{j+1}(2)}{z_i^{j+1}(1)}, \frac{w_{i+1}^j(1)}{w_{i+1}^j(2)}, \frac{w_{i+1}^j(2)}{w_{i+1}^j(1)}\right)$$

$$\leq M^{2P_{i+j(\gamma+1)}(c)}[\{(z_i(l), w^j(l))\}_{l=1}^2]^{\tilde{c}^{i+1+j(\gamma+1)}}.$$

In particular if the initial values are the same:

$$z_i(1) = z_i(2), \quad w^j(1) = w^j(2)$$

then they satisfy the uniform estimates:

$$\max\left(\frac{z_i^{j+1}(1)}{z_i^{j+1}(2)}, \frac{z_i^{j+1}(2)}{z_i^{j+1}(1)}, \frac{w_{i+1}^j(1)}{w_{i+1}^j(2)}, \frac{w_{i+1}^j(2)}{w_{i+1}^j(1)}\right) \leq M^{2P_{i+j(\gamma+1)}(c)}.$$

Notice that for the Mealy case ($\alpha = \beta = 0$), their rates are bounded by:

$$M^{2P_{i+j}(c)}[\{(z_i(l), w^j(l))\}_{l=1}^2]^{\tilde{c}^{i+j+1}}.$$

Proof. Let us consider the solutions to the equations:

$$\begin{cases} x_l'(i, j+1) = \psi_t^l(y_l'(i, j), x_l'(i, j), \ldots, x_l'(i+\alpha, j)) \\ y_l'(i+1, j) = \phi_t^l(y_l'(i, j), x_l'(i, j), \ldots, x_l'(i+\beta, j)) \end{cases}$$

with the initial values $x_l'(i, 0) = \log_t z_i(l)$ and $y_l'(0, j) = \log_t w^j(l)$ for $l = 1, 2$.

With the same initial values, let us also consider the solutions to other equations:

$$\begin{cases} x_l(i, j+1) = \psi^l(y_l(i, j), x_l(i, j), \ldots, x_l(i+\alpha, j)) \\ y_l(i+1, j) = \phi^l(y_l(i, j), x_l(i, j), \ldots, x_l(i+\beta, j)) \end{cases}$$

with $x_l(i,0) = x_l'(i,0)$ and $y_l(0,j) = y_l'(0,j)$.

By proposition 6.2, the estimates hold for $l = 1, 2$:

$$|x_l(i,j+1) - x_l'(i,j+1)|, \ |y_l(i+1,j) - y_l'(i+1,j)| \le P_{i+j(\gamma+1)}(c) \log_t M.$$

On the other hand by lemma 6.5, the estimates hold:

$$|x_1(i,j+1) - x_2(i,j+1)|, \ |y_1(i+1,j) - y_2(i+1,j)|$$
$$\le \tilde{c}^{i+1+j(\gamma+1)} |\{x_i(1), x_i(2)\}_i; \{y^j(1), y^j(2)\}_j|.$$

Thus combining with these, the estimates hold:

$$|x_1'(i,j) - x_2'(i,j)|$$
$$\le |x_1'(i,j) - x_1(i,j)| + |x_1(i,j) - x_2(i,j)| + |x_2(i,j) - x_2'(i,j)|$$
$$\le 2P_{i+(j-1)(\gamma+1)}(c) \log_t M + \tilde{c}^{i+1+(j-1)(\gamma+1)} |\{x_i(1), x_i(2)\}_i; \{y^j(1), y^j(2)\}_j|,$$
$$|y_1'(i,j) - y_2'(i,j)|$$
$$\le |y_1'(i,j) - y_1(i,j)| + |y_1(i,j) - y_2(i,j)| + |y_2(i,j) - y_2'(i,j)|$$
$$\le 2P_{i-1+j(\gamma+1)}(c) \log_t M + \tilde{c}^{i+j(\gamma+1)} |\{x_i(1), x_i(2)\}_i; \{y^j(1), y^j(2)\}_j|.$$

Since $\log_t F^l(i,j) = x_l'(i,j)$ and $\log_t G^l(i,j) = y_l'(i,j)$ hold by lemma 6.6, these estimates verify the conclusions. $\quad\square$

References

Many concrete cases of evolutional discrete dynamics have been studied. This book includes basic subjects in both automata groups in chapter 7 and integrable systems in chapter 10. See references at the end of those chapters. Analysis on rational orbits under tropically equivalent functions has been developed in [Kat8], [Kat9].

Chapter 7

Automata groups

Automata groups are a class in finitely generated groups which contain several instances with quite important and characteristic properties in group theory. It is beyond the scope of this book to give complete references, but let us quickly explain a few important results.

(1) *Intermediate growth groups:* It has been known that many finitely generated groups satisfy an alternative that their growth functions are either exponential or polynomial. Virtually nilpotent groups, solvable groups and hyperbolic groups all satisfy such conditions. Milnor asked whether intermediate growth, such as $\exp(\sqrt{n})$, can exist. The first example of intermediate growth groups was discovered in automata groups by Grigorchuk.

(2) *Burnside groups:* Burnside groups are finitely generated torsion groups which are infinite. So far Adjan–Novikov have presented a Burnside group, but their construction was quite complicated and long. Later Aleshin–Grigorchuk constructed another Burnside group in automata groups. It is given by in a relatively simple way, and later we will outline its construction.

(3) *Classification of automata groups:* A classification program was developed with small numbers of states and alphabets by R. Grigorchuk, V. Nekrashevich and V. Sushchanskii. In the smallest case of 2 states and 2 alphabets, the only new case was the lamplighter group, while other cases turned out to be familiar groups.

In our application of automata groups to tropical geometry, the basic idea is to study how such geometric or analytic properties reflect the structure of the rational dynamics, if they contain such groups as the frame of the dynamical systems. In particular we verify the existence of rational orbits which are *infinite quasi-recursive* in chapter 9.

7.1　Mealy automata

Let Q and A be finite sets called the states and alphabets respectively. Let us consider a Mealy automaton $\mathbf{A}$:

$$\left\{ \begin{array}{l} \psi : Q \times S \to S, \\ \phi : Q \times S \to Q \end{array} \right.$$

which is a special case of the state dynamics with $q_0 = q$:

$$\mathbf{A}_q : (s_0, s_1, \dots) = (s_0', s_1', \dots)$$

defined inductively by:

$$s_i' = \psi(q_i, s_i), \quad q_{i+1} = \phi(q_i, s_i).$$

Lemma 7.1. *A Mealy dynamics induce the level-set actions as:*

$$\mathbf{A}_q : X_S^{N+1} \to X_S^{N+1}$$

where

$$X_S^{N+1} = \{\bar{s}^* = (s_0, s_1, \dots, s_N) : s_i \in S\}$$

are the set of words of length $N + 1$.

This is almost trivial, but it does not hold for general state dynamics if at least one of α or $\beta \geq 1$ holds (see 6.1).

Let $m = \sharp|A|$ be the cardinality of S. The rooted regular m-tree T_m is a tree with a distinguished point $*$ called root such that each vertex is connected with $m + 1$ edges except the root and the root is connected with m edges.

There is a canonical distance on the set of vertices from the root, and let $T_m^d \subset T_m$ be finite subtrees whose vertices have distance less than or equal to d.

Lemma 7.2. *The set of edge of T_m^d is identified with X_S^d. In particular all edges of T_m is identified with $X_S^\infty \equiv \cup_{N \geq 0} X_S^{N+1}$.*

It follows from the above two lemmas that $\mathbf{A}_q$ gives the action on the regular rooted tree:

$$\mathbf{A}_q : (T_m, T_m^d) \to (T_m, T_m^d)$$

which preserves each level set T_m^d. Conversely the action of $\mathbf{A}_q$ is determined on T_m, if it is determined on each level set T_m^d.

Let us say that $\mathbf{A}$ is *invertible*, if

$$\psi(q, \ ) : S \cong S$$

are one to one and onto for all $q \in Q$. An invertibe automaton $\mathbf{A}$ gives automorphisms

$$\mathbf{A}_q : (T_m, T_m^d) \cong (T_m, T_m^d).$$

Definition 7.1. Let $\mathbf{A}$ be an invertible Mealy automaton. The group generated by the set of the action of states:

$$G(\mathbf{A}) = \ \text{gen} \ \{\mathbf{A}_q : T_m \cong T_m : q \in Q\}$$

is called the automata group.

For $X_Q^{L+1} = \{\bar{q}^L = (q^0, \ldots, q^L) : q^i \in Q\}$, the composition:

$$\mathbf{A}_{\bar{q}^L} = \mathbf{A}_{q^L} \circ \ldots \mathbf{A}_{q^0} : T_m \cong T_m$$

is an element in the automata group given by the Mealy automaton.

Remark 7.1. (1) An automata group is a subgroup of $\text{Aut}(T_m)$.

(2) An automata group admits a canonical generating set by states.

(3) General state dynamics give the action

$$\mathbf{A}_q : \partial T_m \to \partial T_m$$

on the boundary of the tree.

(4) In later sections, we embed both S and Q into real numbers, when we study rational orbits corresponding to automata groups.

7.2 Burnside group

Let us outline the construction of the Burnside automata group (refer to [Har] for details).

A key property of automata groups is self-similarity in the following sense. Let $\mathbf{A}$ be an invertible Mealy automaton, and consider the initial automata $\mathbf{A}_{q,i} \equiv \mathbf{A}_{\phi(q,s_i)}$. Since $\psi(q, \ ) : S \cong S$ are the permutations, they can be regarded as elements in the symmetric group $\sigma_q \in S_d$.

Lemma 7.3. *Any automata group G given by a Mealy automaton $\mathbf{A}$ admits the embedding into the wreath product of G with S_d:*

$$\mathbf{A}_q \to (\mathbf{A}_{q,1}, \ldots, \mathbf{A}_{q,d})\sigma_q.$$

Theorem 7.1. *The Mealy automaton diagram in Fig.* 7.1 *represents an infinite torsion group* Γ.

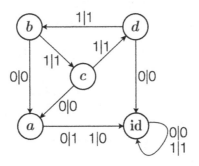

Fig. 7.1 *Diagram of a Mealy automaton*

In general a Mealy automaton can be described by a diagram as in theorem 7.1. The alphabets are $\{0, 1\}$ and the states are $\{a, b, c, d, \text{id}\}$.

$x \mid y$ over the arrow reads x as an input and y as an output. For example $\psi(d, 1) = 1$ and $\psi(d, 0) = 0$, and so on. The arrow represents a change of states. For example $\phi(b, 0) = a$ and $\phi(b, 1) = c$ and so on.

Let us explain how these states trigger the action on T_2. The rule is given in 7.1 above. Let us identify a vertex on T_2 with a $\{0, 1\}$ sequence. Then for example, we give actions $\mathbf{A}_a, \mathbf{A}_b$ and $\mathbf{A}_c$ as follows:

$$\mathbf{A}_a(0011101100000 \cdots) = 1011101100000 \cdots$$
$$\mathbf{A}_b(0011101100000 \cdots) = 0111101100000 \cdots$$
$$\mathbf{A}_c(1001110110000 \cdots) = 1001110110000 \cdots$$

In the case of an invertible Mealy automaton, there is more conventional notation, such as $a = \epsilon$ and $b, c, d = 1$, where ϵ implies the permutation over two alphabets, and hence $\phi(a, x) = \bar{x}$, where:

$$- \; : \{0, 1\} \to \{0, 1\}$$

is the involution given by $\bar{0} = 1$ and $\bar{1} = 0$. 1 implies the identity and hence $\phi(c, \;) = \text{id} : \{0, 1\} \to \{0, 1\}$.

id implies the identity over any alphabet, and hence $\phi(\text{id}, s) = \text{id}$.

Let T_2 be the rooted binary tree and $G \equiv \text{Aut } T_2$ be the automorphism group of T_2. Let $st(1) \subset \text{Aut } T_2$ be the normal subgroup which consists of all elements fixing the first stage of the tree $T_2^1 \subset T_2$. Then one obtains the canonical identification:

$$st(1) \cong G \times G \qquad g \to (g_0, g_1).$$

The automata group in Fig. 7.1 is generated by:

$$\Gamma = \langle\, a, b, c, d \,\rangle$$

since id acts as the identity on T_2.

Let us determine the actions of the generating set on each level set of T_2. Let $(j_1, \ldots, j_{d-1}) \in T_2^d$ where $j_l \in \{0, 1\}$, and denote the action of $\mathbf{A}_a$ by just a, and so on. Notice that:

(1) $a(j_1, j_2, \ldots) = (\bar{j}_1, j_2, \ldots)$ involutes at the first stage and the lower stages are not changed,

(2) $b, c, d \in st(1)$ are recursively defined by:

$$b = (a, c), \quad c = (a, d), \quad d = (1, b).$$

For example, $\phi(b, 0) = a$ and hence

$$b(0, j_2, j_3, \ldots) = (0, a(j_2, j_3, \ldots)) = (0, \bar{j}_2, j_3, \ldots).$$

The following formulas imply that b is defined, as we explain below. The other cases of the states are also similarly defined.

$$b(0, j_2, j_3, \ldots) = (0, \bar{j}_2, j_3, \ldots),$$
$$b(1, j_2, j_3, \ldots) = (1, c(j_2, j_3, \ldots)),$$
$$c(0, j_3, \ldots) = (0, \bar{j}_3, j_4, \ldots),$$
$$c(1, j_3, \ldots) = (0, d(j_3, j_4, \ldots)),$$
$$d(0, j_4, \ldots) = (0, j_4, j_5, \ldots),$$
$$d(1, j_4, \ldots) = (1, b(j_4, j_5, \ldots)).$$

In order to determine the action by b, for example $b(1, 1, 1, j_4, \ldots, j_{d-1}) = (1, 0, 1, b(j_4, \ldots, j_{d-1}))$ and hence b appears again. However the length of $b(j_4, \ldots)$ is strictly smaller than $b(j_1, \ldots)$ and hence one can finally determine the action by b by iterating this process.

The following relations are fundamental:

Lemma 7.4.

$$a^2 = b^2 = c^2 = d^2 = 1,$$
$$bc = cb = d, \quad cd = dc = b, \quad db = bd = c.$$

Proof. They are verified by induction on the length of words and use of recursive formulas. For example,

$$bc(0, j_2, \dots) = b(0, \bar{j}_2, j_3, \dots) = (0, j_2, \dots),$$
$$bc(1, j_2, \dots) = b(1, d(j_2, j_3, \dots)) = (1, cd(j_2, \dots)),$$

and so one can apply the induction hypothesis since the length of $(j_2, \dots)$ is less than $(j_1, \dots)$, and so on. $\square$

In particular, any group element $g \in \Gamma$ is represented by a word of the form:

$$w = u_0 a u_1 a \dots u_{l-1} a u_l$$

where $u_1, \dots, u_{l-1} \in \{b, c, d\}$ and $u_0, u_l \in \{\phi, b, c, d\}$.

Notice that $w \in st(1)$, if and only if a appears even times in the above formula, which is equivalent to that l is even. So $st(1)$ is generated as:

$$st(1) = \langle \ b = (a, c), \ c = (a, d),$$
$$d = (1, b), \ aba = (c, a), aca = (d, a), ada = (b, 1) \ \rangle$$

Corollary 7.1. Γ *is infinite.*

Proof. Let

$$\phi = (\phi_0, \phi_1) : st(1) \to \Gamma \times \Gamma$$

be the above homomorphism. Then each ϕ_i is surjective since their images contain $\{a, b, c, d\}$ by the above. It can happen only when Γ is infinite, since $st(1) \subset \Gamma$ is a proper subgroup. $\square$

Lemma 7.5. *(1)* $\langle \ a, d \ \rangle$ *is order 8 group.*
    (2) $\langle \ a, c \ \rangle$ *is order 16 group.*
    (3) $\langle \ a, b \ \rangle$ *is order 32 group.*

Proof. (1) $(ad)^2 = adad = (b, 1)(1, b) = (b, b)$, and so $(ad)^4 = 1$.
    (2) $(ac)^2 = (da, ad)$.
    (3) $(ab)^2 = (ca, ac)$. $\square$

Proof of theorem 7.1.

We verify that for any $g \in \Gamma$, there is some N with $g^{2^N} = 1$. We will use invariance of the property under change by conjugation in many places below.

Suppose g is represented by a word w of length k.

$$g = 1 \;\Leftarrow\; k = 0$$
$$g^2 = 1 \;\Leftarrow\; k = 1$$
$$g^{32} = 1 \;\Leftarrow\; k = 2.$$

Let us proceed by induction on k. Assume it holds for all words of length up to $k - 1$.

Case 1: Suppose $k = 2l - 1$ is odd. Then:

$$w = \begin{cases} au_1 \ldots u_{l-1}a & \text{or} \\ u_1 a \ldots au_{l-1} \end{cases}$$

If $w = au_1 \ldots u_{l-1}a$, then conjugation by a changes it to $awa = u_1 \ldots u_{l-1}$, whose length is $k - 2$, for which the claim holds by the induction hypothesis.

If $w = u_1 \ldots u_{l-1}$, then $u_1 w u_1 = u_2 \ldots u_{l-2}u_{l-1}u_1$, where $u_{l-1}u_1 \in \{b, c, d\}$ by lemma 7.4. So the length is $k - 1$ and the claim holds.

Case 2: Suppose $k = 2l$ is even. The one may assume:

$$w = au_1 \ldots au_l$$

by taking conjugation by an element in $\{b, c, d\}$, if necessary.

If l is even, then $g \in st(1)$, and

$$\psi(g) = \psi(au_1 a)\psi(u_2) \ldots \psi(au_{2m-1}a)\psi(u_{2m}) \equiv (g_0, g_1)$$

where the lengths of both g_0, g_1 are less than $\frac{k}{2}$, and so the claim holds.

If l is odd, then $k = 4m - 2$, and consider g^2:

$$ww = (au_1 a)u_2 \ldots u_{2m-2}(au_{2m-1}a)u_1(au_2 a) \ldots (au_{2m-2}a)u_{2m-1}.$$

Notice that both $au_i a$ and u_i appear for all $1 \le i \le 2m - 1$ in the above product.

$$\psi(g^2) = \psi(au_1 a)\psi(u_2) \cdots \equiv (\alpha, \beta)$$

where the lengths of both α, β are less than $4m - 2$.

If $u_j = d$ for some j, then:

$$\psi(au_j a) = (b, 1), \quad \psi(u_j) = (1, b)$$

and so the lengths of both α, β are actually less than $4m - 3$, and so the claim holds.

If $u_j = c$ for some j, then:

$$\psi(au_j a) = (d, a), \quad \psi(u_j) = (a, d)$$

and so the lengths of both α, β are less than $4m - 2$ and both contain d. So the claim holds.

Finally If $u_j = b$ for some j, then $\psi(au_j a) = (c, a)$, $\quad \psi(u_j) = (a, c)$, and so the lengths of both α, β are less than $4m - 2$ and both contain c. So the claim holds.

This finishes all the cases.

7.3 Lamplighter group

Let A and B be two groups and suppose A acts on a space X. Let $B_X = \{f : X \to B\}$ be a set of functions, and consider the induced action of A on B_X by:

$$(af)(x) = f(a^{-1}x).$$

We assume that B_X is invariant under the action by A.

Then one obtains the semi-direct product group $B_X \rtimes A$ which is $B_X \times A$ as a set, and the multiplication is given by:

$$(f, a)(f', a') = (f(af'), aa').$$

The lamplighter group is a finitely generated infinitely presented group:

$$(\oplus_{\mathbb{Z}} \mathbb{Z}_2) \rtimes \mathbb{Z}$$

where $A = \mathbb{Z}$, $B = \mathbb{Z}_2$ with $X = \mathbb{Z}$, and the action of A on X is the translation, where B_X is the set of compactly supported function on $\mathbb{Z}$.

The group admits the canonical two generators:

$$v = (\ldots, 0, 1, 0, \ldots) \in \oplus_{\mathbb{Z}} \mathbb{Z}_2, \quad u = 1 \in \mathbb{Z}.$$

Theorem 7.2. *The automata group corresponding to the Mealy automaton as given in the diagram in Fig. 7.2.*

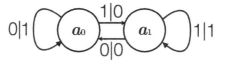

Fig. 7.2

is isomorphic to the lamplighter group whose generating sets admit corre-spondences:

$$(a_0)^{-1}a_1 = v, \qquad a_0 = u.$$

Let us just re-express the diagram in Fig. 7.2.

$$\psi : \{a^0, a^1\} \times \{s_0, s_1\} \to \{s_0, s_1\}, \quad \psi(a^0, \ ) = \epsilon, \ \psi(a^1, \ ) = \mathrm{id},$$
$$\phi : \{a^0, a^1\} \times \{s_0, s_1\} \to \{a^0, a^1\}, \quad \phi(\ , s_i) = a^i \ (i = 0, 1)$$

as the pair of functions.

Below we give an outline of the proof following [GZ]. For this we need some preliminary notions as below.

Let $v \in T_d$ be a vertex. We say that an action of G on T_d is *self-similar*, if the group isomorphism $H \equiv st_G(v) \cong G$ holds, after identification of T_d with T_v, where $T_v \subset T_d$ is the maximal connected subtree with the root v.

Any automata group satisfies self-similarity.

It is quite useful to understand elements in G in a geometric way, which is called a *portrait*. Let $g \in G$. For each vertex $v \in T_d$, there is an associated permutation $g(v) \in S_d$. The set:

$$\{g(v) : v \in T_d\}$$

is called the portrait of g.

The portrait uniquely determines the action and hence the element $g \in G$.

Portrait has depth m, if it is the smallest integer such that each vertex of level more than or equal to $m + 1$ is identity.

Proof of theorem 7.2.

Step 1: Let $\gamma = a^{-1}b$ and $\alpha = ab^{-1}$. G is generated by γ and a. The strategy is to proceed as follows:

(1) a map:

$$\oplus_{\mathbb{Z}} \mathbb{Z}_2 \to N' = \{\gamma^{a^{n_1}} \dots \gamma^{a^{n_s}}\}_{n_t \in \mathbb{Z}} \subset G$$

by $(1_{n_1}, \dots, 1_{n_s}) \to \gamma^{a^{n_1}} \dots \gamma^{a^{n_s}}$ gives an isomorphism between the free abelian groups.

(2) N' is normal, and G/N' is free cyclic.

(3) Construct a group $\bar{N} \subset G$ whose element has order 2.

(4) There is a group embedding $N' \hookrightarrow \bar{N}$.

Step 2: Let us complete the proof of theorem 7.2 under assumptions (1) to (4). The assumptions give an isomorphism from the lamplighter group to the automata group of the lamplighter automaton as follows: since any element has order 2, $N' \cong \oplus_{\mathbb{Z}} \mathbb{Z}_2$. G is generated by a and γ, and G/N is infinite cyclic generated by a. The multiplication by a induces the shift on N. For example, let us check:

$$a\gamma^{a^l} = aa^{-l}\gamma a^l = a^{-(l-1)}\gamma a^{l-1}a = \gamma^{a^{l-1}}a.$$

Hence we obtain the isomorphisms between groups:

$$G \cong (\oplus_{\mathbb{Z}} \mathbb{Z}_2) \rtimes \mathbb{Z}.$$

This gives the conclusion.

Step 3: We have the embedding from G to the wreath product of G with S_2:

$$\theta(x) = \begin{cases} (a,b)\epsilon & x = a \\ (a,b) & x = b. \end{cases}$$

In particular the depths of both a and b are infinite.

We have useful formulas:

$$\gamma = (b^{-1}, a^{-1})\epsilon(a,b) = (1,1)\epsilon,$$
$$\alpha^2 = ab^{-1}ab^{-1} = (a,b)\epsilon(a^{-1},b^{-1})(a,b)\epsilon(a^{-1},b^{-1}) = 1,$$
$$\alpha = (a,b)\epsilon(a^{-1},b^{-1}) = (ab^{-1},ba^{-1})\epsilon = (\alpha,\alpha^{-1})\epsilon = (\alpha,\alpha)\epsilon.$$

Step 4: Let us introduce another subset $\bar{N} \subset G$. An element $h \in \bar{N}$, if there exists an infinite sequence $\{h_i\}_i$ with $h = h_1$ such that

$$h_i = (h_{i+1}, h_{i+1}) \text{ or } (h_{i+1}, h_{i+1})\epsilon$$

Notice that $h_i \in \bar{N}$ holds by definition. Moreover $\gamma, \alpha \in \bar{N}$ holds by the above computations.

Assertion 7.1. $\bar{N}$ is a group. α, γ and any element in $\bar{N}$ commute each other.

Moreover any element $h \in \bar{N}$ has order 2.

Proof. It is straightforward to check that $\bar{N}$ is a group.

We have the formulas:

$$\gamma h = \begin{cases} (1,1)\epsilon(h',h') = (h',h')\epsilon & \text{if } h = (h',h') \\ (1,1)\epsilon(h',h')\epsilon = (h',h') & \text{if } h = (h',h')\epsilon \end{cases}$$

$$h\gamma = \begin{cases} (h',h')(1,1)\epsilon = (h',h')\epsilon & \text{if } h = (h',h') \\ (h',h')\epsilon(1,1)\epsilon = (h',h') & \text{if } h = (h',h')\epsilon. \end{cases}$$

So h and γ are commutative.

Similarly, we have the formulas:

$$\begin{cases} \alpha h = (\alpha h', \alpha h')\epsilon, \quad h\alpha = (h'\alpha, h'\alpha)\epsilon & \text{if } h = (h',h') \\ \alpha h = (\alpha h', \alpha h'), \quad h\alpha = (h'\alpha, h'\alpha) & \text{if } h = (h',h')\epsilon. \end{cases}$$

These formulas imply that $h\alpha$ and αh have the same portrait mutually. This verifies (1).

Since $h_i^2 = (h_{i+1}^2, h_{i+1}^2)$, the portraits consist of only 1.

$\square$

(3) follows from this sub lemma.

Step 5: We check that the depth of the portrait of $\gamma_n \equiv \gamma^{a^n}$ is n. Actually the depth of γ is 0. For example:

$$a^{-1}\gamma a = (b^{-1}, a^{-1})\epsilon(1,1)\epsilon(a,b)\epsilon = (b^{-1}, a^{-1})\epsilon(b,a)$$
$$= (b^{-1}a, a^{-1}b)\epsilon = (\gamma^{-1}, \gamma)\epsilon = (\gamma, \gamma)\epsilon,$$
$$b^{-1}\gamma b = (a^{-1}, b^{-1})(1,1)\epsilon(a,b) = (\gamma, \gamma^{-1})\epsilon = (\gamma, \gamma)\epsilon,$$
$$a^{-2}\gamma a^2 = (b^{-1}, a^{-1})\epsilon(\gamma^{-1}, \gamma)(a,b)\epsilon$$
$$= (b^{-1}, a^{-1})(\gamma, \gamma^{-1})(b,a) = (b^{-1}\gamma b, a^{-1}\gamma^{-1}a) = (b^{-1}\gamma b, a^{-1}\gamma a)$$

and their depths are $1, 1$ and 2 respectively. One can proceed by induction.

The above computation applied to the other cases such as $a\gamma a^{-1}$ etc., plus assertion 7.1 gives the embedding $N' \hookrightarrow \bar{N}$, and verifies (4) above. Moreover any two elements in N' commute each other.

Step 6: Let us verify (1) above. It remains to check that there are no relations among elements in N'. $\{\gamma_n\}_n$ are linearly independent. Otherwise, there is some:

$$\gamma^{a^{j_1}} \cdots \gamma^{a^{j_s}} = 1$$

which holds for some $j_1 < j_2 \cdots < j_s$. Then we must have the equality:

$$\gamma^{a^{j_s}} = \gamma^{a^{j_1}} \cdots \gamma^{a^{j_{s-1}}}$$

which is impossible, since their depths are different between the right and left hand sides.

In particular a is infinite cyclic.

Step 7: Let us verify (2) above. Surely N' is normal. G/N' cannot be trivial, since a has ∞ depth. So G/N' is infinite cyclic generated by a.

7.4 Automata twisted groups and their classification

Let us try to pick up Mealy automata up to equivalence, which satisfy the following four properties: (1) conservativity in the sense preserving the number of 1; (2) soliton characteristics in the sense that speed of solitary sequences does not change after interaction; (3) twisted invertiblity; (4) uniqueness of the final state in the sense that there is a k such that any state can change into a specified state if 0 appears more than k times.

According to Tsujimoto's computation using a computer algebra system, the results are rather surprising, since most of the automata have been ruled out and quite a small number of automata can survive and hold these properties.

states	results	possible candidates
2	BBS$_2$	$4^4 = 256$ cases
3	BBS$_3$+ other two cases	$6^6 = 46656$ cases
4	BBS$_4$+ less than 6 cases	$8^8 \sim 16$ million cases
5	BBS$_5$+ less than 13 cases	$10^{10} \sim 10$ billion cases
6	BBS$_6$+ less than 14 cases	$12^{12} \sim 8.9$ trillion cases

Here we have omitted a trivial case of the identity action.

At present we have drastically decreased the number of possible candidates, but it is not clear how many automata precisely satisfy these properties, when the number of the states is larger than or equal to 3. BBS$_k$ satisfy these three properties for all $k \geq 2$, and hence the classifications contain non empty sets for any number of the states.

In the case of 3 states, the other two cases have been checked to satisfy these properties. Actually they consist of two new integrable automata described by the two diagrams in Fig. 7.3.

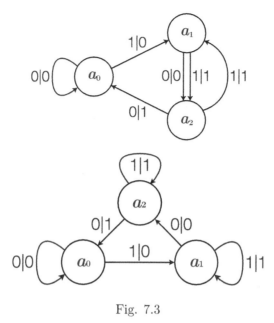

Fig. 7.3

References

Burnside groups have been constructed in [AN], [Ale], [Grig1]. We followed the construction of Burnside group in [Har], and the construction of lamplighter automaton in [GZ]. The intermediate growth group was given in [Grig2], see also [Har]. Brieussel [Bri] found automata groups whose growth exponents fluctuate. Automata groups with non uniform exponential growth have been given in [Wil]. The classification of automata groups with a small number of states was given in [GNS]. The method in 7.4 uses a quite different approach from it. A comprehensive foundation of geometric group theory is given in [Gro2].

Chapter 8

Stable state dynamics

Automaton is defined over finite sets, but once it is expressed by $(\max, +)$-functions, it canonically extends to a map over the real vector space. If one starts from an automaton, there are no canonical extensions, which are actually so flexible. Here we verify that there still exist effective extensions to reflect dynamical properties of automata to rational dynamics passing through tropical geometry.

8.1 Extensions of automata

Let $Q, S \subset \mathbb{R}$ be finite sets, and take an automaton

$$
\begin{cases}
\psi : Q \times S^{\alpha+1} \to S, \\
\phi : Q \times S^{\beta+1} \to Q.
\end{cases}
$$

Let:

$$
\begin{cases}
\tilde{\psi} : \mathbb{R} \times \mathbb{R}^{\alpha+1} \to \mathbb{R}, \\
\tilde{\phi} : \mathbb{R} \times \mathbb{R}^{\beta+1} \to \mathbb{R}
\end{cases}
$$

be two maps.

Let us say that the pair of functions $(\tilde{\psi}, \tilde{\phi})$ extends the automaton, if their restrictions coincide with each other:

$$
\begin{cases}
\tilde{\psi}|Q \times S^{\alpha+1} = \psi, \\
\tilde{\phi}|Q \times S^{\beta+1} = \phi.
\end{cases}
$$

Conversely if there are finite subsets $Q, S \subset \mathbb{R}$ so that their restrictions of the pair of functions $(\tilde{\psi}, \tilde{\phi})$ induce the functions:

$$
\begin{cases}
\psi \equiv \tilde{\psi} : Q \times S^{\alpha+1} \to S, \\
\phi \equiv \tilde{\phi} : Q \times S^{\beta+1} \to Q
\end{cases}
$$

then we say that $(\tilde{\psi}, \tilde{\phi})$ restrict to an automaton (ψ, ϕ).

Notice that if a pair of bounded functions restricts over the integer as:

$$\begin{cases} \tilde{\psi} : \mathbb{Z} \times \mathbb{Z}^{\alpha+1} \to \mathbb{Z}, \\ \tilde{\phi} : \mathbb{Z} \times \mathbb{Z}^{\beta+1} \to \mathbb{Z} \end{cases}$$

then they restrict to some automaton for some $Q, S \subset \mathbb{Z}$. In fact one can put $Q = S = \text{im } \tilde{\psi} \cup \text{im } \tilde{\phi}$.

8.2   Extensions by relative (max, +)-functions

Let $\varphi : \mathbb{Z}^n \to \mathbb{Z}$ be a map. There are two canonical ways of the extensions:

(1) By connecting these integer values by segments, one can extend φ straightforwardly to a piecewisely linear map:

$$\varphi : \mathbb{R}^n \to \mathbb{R}.$$

(2) If $\varphi : \mathbb{Z}^n \to \mathbb{Z}$ is equipped with its presentation by a (max, +)-function, then it is canonically extended over the real numbers.

For our purposes we will choose the method (2) above.

Lemma 8.1. *Let us consider a Mealy automaton* $\psi : Q \times S \to S$, $\phi : Q \times S \to Q$ *with* $Q, S \subset \mathbb{R}$. *Then there is a bounded extension:*

$$\begin{cases} \tilde{\psi} : \mathbb{R} \times \mathbb{R} \to \mathbb{R}, \\ \tilde{\phi} : \mathbb{R} \times \mathbb{R} \to \mathbb{R} \end{cases}$$

so that both functions $\tilde{\psi}$ *and* $\tilde{\phi}$ *are represented by relative* (max, +)-*functions.*

Proof. It is easy to replace functions by bounded ones. In fact if a function $\tilde{\varphi} : \mathbb{R}^2 \to \mathbb{R}$ is unbounded, then let us put $M = \max \tilde{\varphi}(Q \times S)$ and $m = \inf \tilde{\varphi}(Q \times S)$. $\tilde{\varphi}' = \max(m, \tilde{\varphi})$ still remains an extension of $\varphi = \tilde{\varphi}|Q \times S$, and $\tilde{\varphi}'' = \inf(\tilde{\varphi}', M) = -\max(-\tilde{\varphi}', -M)$ is the desired one.

Step 1: We extend maps inductively on the number of the points of $Q \times S$.

Let us describe how to extend a map on neighborhood of a triangle. Let $\{m_1, m_2, m_3\} \subset Q \times S$ be three points which are mutually different, and consider the triangle $\Delta \subset \mathbb{R}^3$ whose vertices are given by the graph:

$$O_i = (m_i, \psi(m_i)) \in \mathbb{R}^3 \qquad (i = 1, 2, 3).$$

Let $l_i \subset \Delta$ be the boundary segments whose end points are O_i and O_{i+1} mod 3.

Let us choose planes L_i which contain l_i so that they are represented by graphs of affine linear functions $\varphi_i : \mathbb{R}^2 \to \mathbb{R}$ as

$$L_i = \{(x_1, x_2, \varphi_i(x_1, x_2)) : (x_1, x_2) \in \mathbb{R}^2\}.$$

Let $g \in \Delta$ be the barycenter, and $l : [0, \infty)$ be the half lines starting from g which are vertical to $\mathbb{R}^2$ in $\mathbb{R}^3 = \mathbb{R}^2 \times \mathbb{R}$. There are just two choices l_b and l_a which go below or above Δ.

For $i = 1, 2, 3$, let L_i^b and L_i^a be other planes which contain l_i and intersect with l_b and l_a at $l_b(t)$ or $l_a(t)$ respectively. So three planes $\{L_1^b, L_2^b, L_3^b\}$ intersect at one point $l_b(t)$. Similarly for a. We will choose these planes with sufficiently large $t >> 1$. Let φ_i^b and φ_i^a be their representations by affine linear functions.

Now we have two types of the cones:

$$C_b = \text{ graph } \max(\varphi_1^b, \varphi_2^b, \varphi_3^b)$$

and C_a are similar. C_b is concave and C_a is convex. Notice that we may choose arbitrarily sharp slopes of the cones.

Let us choose a large C with $C \geq \max\{|\psi(m)| : m \in Q \times S\}$. Let φ be another affine linear function whose graph contains Δ.

Now we put the bounded functions:

$$\psi_\Delta^b = -\max(-C, -\max(\varphi, \varphi_1^b, \varphi_2^b, \varphi_3^b)),$$
$$\psi_\Delta^a = \max(-C, -\max(-\varphi, -\varphi_1^a, -\varphi_2^a, -\varphi_3^a)).$$

Notice that both graphs contain Δ.

This is a bounded extension over three points.

Step 2: Let us order all the points $\{m_{-1}, m_0, m_1, m_2, \ldots, m_l\} = Q \times S$, so that there are families of two dimensional polytopes M_i such that:

(1) M_i is given by unions of M_{i-1} with a single triangle Δ_i which contain the vertices m_i and
(2) M_i does not contain the sets $\{m_{i+1}, \ldots, m_l\}$.

Now we inductively construct relative $(\max, +)$-functions $\tilde{\psi}_i : \mathbb{R}^2 \to \mathbb{R}$ so that the equalities:

$$\tilde{\psi}_i(m_k) = \psi(m_k)$$

hold for all $1 \leq k \leq i$.

We have constructed $\psi_1^a = \psi_{\Delta_1}^a$ by step 1, with $\Delta_1 = \Delta$.

Suppose we have obtained $\tilde{\psi}_i$, let us construct $\tilde{\psi}_{i+1}$. Divide into two cases:

Suppose $\tilde{\psi}_i(m_{i+1}) \geq \psi(m_{i+1})$. Let $\psi^b_{\Delta_{i+1}}$ be as in step 1. By choosing sufficiently sharp slopes, we may assume that the estimates:

$$\psi^b_{\Delta_{i+1}}(m_k) \geq \tilde{\psi}_i(m_k)$$

hold for all $1 \leq k \leq i$. Then we put the bounded function as:

$$\tilde{\psi}_{i+1} = -\max(-\psi^b_{\Delta_{i+1}}, -\tilde{\psi}_i).$$

Next suppose $\tilde{\psi}_i(m_{i+1}) \leq \psi(m_{i+1})$. Let $\psi^a_{\Delta_{i+1}}$ be as in step 1. Again by choosing sufficiently sharp slopes, we may assume that the estimates:

$$\psi^a_{\Delta_{i+1}}(m_k) \leq \tilde{\psi}_i(m_k)$$

hold for all $1 \leq k \leq i$. Then we put the bounded function as:

$$\tilde{\psi}_{i+1} = \max(\psi^a_{\Delta_{i+1}}, \tilde{\psi}_i).$$

In any case $\tilde{\psi}_{i+1}$ above satisfies the desired properties. This completes the induction step. $\qquad\qquad\square$

8.3   Stability and extensions

Let us introduce stability under iterations of maps and study their long-time behaviors.

Let $Q, S \subset \mathbb{R}$ be finite sets, and consider a map $\varphi : Q \times S^a \to S$ with its extension $\varphi : \mathbb{R} \times \mathbb{R}^a \to \mathbb{R}$.

Definition 8.1. $\varphi : \mathbb{R} \times \mathbb{R}^a \to \mathbb{R}$ is stable over (Q, S), if there are $0 < \delta < 1$ and $0 \leq \mu < 1$ so that the estimates hold:

$$|\varphi(y, \bar{x}) - \varphi(q, \bar{s})| \leq \mu \, d((y, \bar{x}), (q, \bar{s}))$$

for any $(y, \bar{x}) \in \mathbb{R} \times \mathbb{R}^a$ and $(q, \bar{s}) \in Q \times S^a$ with $d((y, \bar{x}), (q, \bar{s})) < \delta$.

Notice that φ is (δ', μ)-stable for all $0 < \delta' \leq \delta$, whenever it is (δ, μ)-stable.

Remark 8.1. The above inequality by a positive constant less than 1 is called a contracting property. It would not be realistic to use contracting maps everywhere, since such classes of maps are too small. That is why we use partially contracting maps as above.

Let **A** be an automaton given by a pair of two functions

$$\begin{cases} \psi : Q \times S^{\alpha+1} \to S, \\ \phi : Q \times S^{\beta+1} \to Q. \end{cases}$$

A *stable extension* of **A** is given by two stable extensions of the functions:

$$\begin{cases} \psi : \mathbb{R} \times \mathbb{R}^{\alpha+1} \to \mathbb{R}, \\ \varphi : \mathbb{R} \times \mathbb{R}^{\beta+1} \to \mathbb{R}. \end{cases}$$

Example 8.1. (1) $\psi : \mathbb{R} \times \mathbb{R}^{\alpha+1} \to \mathbb{R}$ is a stable extension of $\psi : Q \times S^{\alpha+1} \to S$ with $\mu = 0$, if it is locally constant in small neighborhoods of each $(q, \bar{s}) \in Q \times S^{\alpha+1}$.

(2) $\psi : \mathbb{R} \times \mathbb{R} \to \mathbb{R}$ is a stable extension of $\psi : Q \times S \to S$, if it is of the form with $\max(|\alpha_{q,s}|, |\beta_{q,s}|) < 1$:

$$\psi(y, x) = \alpha_{q,s}(x - s) + \beta_{q,s}(y - q) + \psi(q, s)$$

in small neighborhood of each $(q, s) \in Q \times S$.

Proposition 8.1. *Let $Q, S \subset \mathbb{R}$ be finite sets and (ψ, ϕ) be a Mealy automaton $\psi : Q \times S \to S$ and $\phi : Q \times S \to S$.*

Then there is a bounded and stable extension of the automaton:

$$\begin{cases} \tilde{\psi} : \mathbb{R} \times \mathbb{R} \to \mathbb{R}, \\ \tilde{\phi} : \mathbb{R} \times \mathbb{R} \to \mathbb{R} \end{cases}$$

so that both $\tilde{\psi}$ and $\tilde{\phi}$ are represented by relative $(\max, +)$-functions.

Proof. The proof consists of a minor modification of lemma 8.1.

Let us choose disjoint union of squares in $\mathbb{R}^2$ so that each square contains exactly one point $(q, s) \in Q \times S$ in its interior. Denote such a square by $D(q, s) \subset \mathbb{R}^2$.

Firstly let us put locally constant functions as:

$$\tilde{\psi}(y, x) = \psi(q, s), \quad \tilde{\phi}(y, x) = \phi(q, s)$$

for $(x, y) \in D(q, s)$. This determines the functions $\tilde{\psi}$ and $\tilde{\phi}$ in a small neighborhood of $Q \times S \subset \cup_{(q,s) \in Q \times S} D(q, s)$.

For the rest, one can follow the same argument as the proof of lemma 8.1, and extend the domains of these functions inductively. □

8.4 Stable state dynamics

Let $Q, S \subset \mathbb{R}$ be finite sets, and consider an automaton $\mathbf{A}$ defined by $\psi : Q \times S^{\alpha+1} \to S$ and $\phi : Q \times S^{\beta+1} \to Q$. Recall state dynamics in 6.3. For each $q \in Q$ and $q_0 = q$, let us consider:

$$\mathbf{A}_q : X_S \to X_S = \{ (k_0, k_1, \dots) : k_i \in S \}$$
$$\mathbf{A}_q(k_0, k_1, \dots) = (k_0', k_1', \dots),$$
$$\begin{cases} k_i' = \psi(q_i, k_i, \dots, k_{\alpha+1}), \\ q_{i+1} = \phi(q_i, k_i, \dots, k_{\beta+1}). \end{cases}$$

For a sequence $\bar{q}^l = (q^0, q^1, \dots, q^l)$, let us denote the compositions of the dynamics $\mathbf{A}_{\bar{q}^l} \equiv \mathbf{A}_{q^l} \circ \cdots \circ \mathbf{A}_{q^0} : X_S \to X_S$.

Let us introduce the space of the real sequences:

$$X_{\mathbb{R}} = \{ \bar{x} = (x_0, x_1, \dots) : x_i \in \mathbb{R} \}$$

and equip the uniform distance between two elements $\bar{x} = (x_0, x_1, \dots)$ and $\bar{x}' = (x_0', x_1', \dots) \in X_{\mathbb{R}}$ with:

$$d(\bar{x}, \bar{x}') = \sup_{0 \le i < \infty} |x_i - x_i'| \in [0, \infty].$$

Let:

$$\begin{cases} \psi : \mathbb{R} \times \mathbb{R}^{\alpha+1} \to \mathbb{R}, \\ \phi : \mathbb{R} \times \mathbb{R}^{\beta+1} \to \mathbb{R} \end{cases}$$

be a stable extension of $\mathbf{A}$ with the constants (δ, μ). For each $y \in \mathbb{R}$ and $y_0 = y$, let us consider the dynamics with the same rule:

$$\mathbf{A}_y : X_{\mathbb{R}} \to X_{\mathbb{R}}$$
$$\mathbf{A}_y(x_0, x_1, \dots) = (x_0', x_1', \dots),$$
$$\begin{cases} x_i' = \psi(y_i, x_i, \dots, x_{\alpha+1}), \\ y_{i+1} = \phi(y_i, x_i, \dots, x_{\beta+1}). \end{cases}$$

For $\bar{y}^l = (y^0, y^1, \dots, y^l) \in \mathbb{R}^{l+1}$, we let:

$$\mathbf{A}_{\bar{y}^l} \equiv \mathbf{A}_{y^l} \circ \cdots \circ \mathbf{A}_{y^0} : X_{\mathbb{R}} \to X_{\mathbb{R}}.$$

This is exactly the state dynamics in 6.3.

For finite sequences $\bar{q}^l \in X_Q^{l+1}$ and $\bar{y}^l \in X_{\mathbb{R}}^{l+1}$, we also equip the same uniform norm with $d(\bar{q}^l, \bar{y}^l) = \sup_{0 \le i \le l} |q^j - y^j| \in [0, \infty)$.

Lemma 8.2. *Let us choose pairs of the sequences, $\bar{y}^l \in \mathbb{R}^{l+1}$ with $\bar{q}^l \in X_Q^{l+1}$, and $\bar{x} \in X_\mathbb{R}$ with $\bar{k} \in X_S$, so that they have bounded distances by $0 < \delta < 1$ from each other:*

$$d(\bar{x}, \bar{k}), \quad d(\bar{y}, \bar{q}) \quad < \quad \delta.$$

Then the estimate holds:

$$d(\mathbf{A}_{\bar{q}}(\bar{k}), \mathbf{A}_{\bar{y}}(\bar{x})) \quad < \quad \delta.$$

Proof. We split the proof into two steps.

Step 1: Firstly let us consider the case $l = 0$, and put $y_0 = y^0 (= y_0^0)$. $|q^0 - y^0| < \delta$ holds by assumption. So let us verify the estimates:

$$|q_{i+1} - y_{i+1}|, \quad |k_i' - x_i'| \; < \delta$$

by induction on $i = 0, 1, 2, \ldots$

Let us start from the former case, and suppose they hold up to i. Then we have the estimates:

$$\begin{aligned}
|q_{i+1} - y_{i+1}| &= |\phi(q_i, k_i, \ldots, k_{\beta+i}) - \phi(y_i, x_i, \ldots, k_{\beta+i})| \\
&\leq \mu \max_{i \leq j \leq \beta+i} \{|q_i - y_i|, |k_j - x_j|\} < \mu\delta < \delta.
\end{aligned}$$

So it holds also at $i + 1$. Thus $|q_i - y_i| < \delta$ holds for all i by induction.

For the latter case also, we have the desired estimates:

$$\begin{aligned}
|k_i' - x_i'| &= |\psi(q_i, k_i, \ldots, k_{\alpha+i}) - \psi(y_i, x_i, \ldots, k_{\alpha+i})| \\
&\leq \mu \max_{i \leq j \leq \alpha+i} (|q_i - y_i|, |k_j - x_j|) < \delta
\end{aligned}$$

by the same way. Thus we obtain the bounds:

$$d(\mathbf{A}_{q^0}(\bar{k}), \mathbf{A}_{y^0}(\bar{x})) \quad < \delta.$$

Step 2: Let us replace the pairs $(\bar{k}, \bar{x})$ by $(\mathbf{A}_{q^0}(\bar{k}), \mathbf{A}_{y^0}(\bar{x}))$ and (q^0, y^0) by (q^1, y^1) in step 1. We apply the same process and obtain the estimates:

$$d(\mathbf{A}_{q^1}(\mathbf{A}_{q^0}(\bar{k})), \mathbf{A}_{y^1}(\mathbf{A}_{y^0}(\bar{x}))) \quad < \quad \delta.$$

By iterating this process l times, one obtains the conclusion.

$\square$

8.5   Change of automata structure

For $l = 1, 2$, let:

$$\mathbf{A}_l : \left\{ \begin{array}{l} \psi^l : Q \times S^{\alpha+1} \to S, \\ \phi^l : Q \times S^{\beta+1} \to Q \end{array} \right.$$

be two automata. For each $q \in Q$, let $(\mathbf{A}_l)_q : X_S \to X_S$ be their dynamics and denote:

$$(\mathbf{A}_l)_q(k_0, k_1, \dots) = (k_0'(l), k_1'(l), \dots)$$

where $k_i'(l)$ is inductively determined by:

$$k_i'(l) = \psi^l(q_i(l), k_i(l), \dots, k_{\alpha+i}(l)), \quad q_{i+1}(l) = \phi^l(q_i(l), k_i(l), \dots, k_{\beta+i}(l))$$

with $q_0(l) = q$.

It can happen that two different automata $\mathbf{A}_1$ and $\mathbf{A}_2$ give the same dynamics. In such a situation, the equalities always hold:

$$(k_0'(1), k_1'(1), \dots) = (k_0'(2), k_1'(2), \dots)$$

while the state sequences may differ from each other:

$$(q^0(1), q^1(1), \dots) \neq (q^0(2), q^1(2), \dots).$$

Let us recall that $\mathbf{A}_1$ and $\mathbf{A}_2$ are called equivalent, if they give the same dynamics as above (definition 6.3).

Definition 8.2. Let $R \subset Q$ be a subset. If the equalities $(\mathbf{A}_1)_q = (\mathbf{A}_2)_q$ hold for all $q \in R$, then we say that $\mathbf{A}_1$ and $\mathbf{A}_2$ are equivalent over R.

Let us give stable extensions of $\mathbf{A}_1$ and $\mathbf{A}_2$ by relative $(\max, +)$-functions (ϕ^1, ψ^1) and (ϕ^2, ψ^2) respectively with the constants (δ, μ).

Let us consider the corresponding state dynamics:

$$\left\{ \begin{array}{l} x_l(i, j+1) = \psi^l(y_l(i, j), x_l(i, j), \dots, x_l(i+\alpha, j)), \\ y_l(i+1, j) = \phi^l(y_l(i, j), x_l(i, j), \dots, x_l(i+\beta, j)) \end{array} \right.$$

with the initial values $\bar{x}(l) = \{x_i(l)\}_i$ and $\bar{y}(l) = \{y^j(l)\}_j$ respectively. For $l = 1, 2$, let us denote:

$$\bar{x}^j(l) = (x_l(0, j), x_l(1, j), \dots) \in X_{\mathbb{R}}.$$

For an infinite sequence $\bar{q} = (q^0, q^1, \dots) \in X_Q$, we denote its restrictions as $\bar{q}^l = (q^0, q^1, \dots, q^l) \in X_Q^{l+1}$.

Notice the equalities $\bar{x}^j(l) = (\mathbf{A}_l)_{\bar{y}^{j-1}}(\bar{x}(l))$.

Corollary 8.1. *Suppose the following conditions:*

(1) There exists a subset $R \subset Q$ so that $\mathbf{A}_1$ and $\mathbf{A}_2$ are equivalent over R.

(2) The initial data satisfy the uniform estimates:

$$d(\bar{x}(l), \bar{k}), \quad d(\bar{y}(l), \bar{q}) \ < \ \delta$$

for some $\bar{k} \in X_S$ and $\bar{q} \in X_R$ for $l = 1, 2$.

Then the uniform estimates hold for all $0 \le j < \infty$:

$$d(\ \bar{x}^j(1), \ \bar{x}^j(2) \ ) \ < \ 2\delta.$$

Proof. The condition (1) implies that the equalities:

$$(\mathbf{A}_1)_{\bar{q}^l} = (\mathbf{A}_2)_{\bar{q}^l} : X_S \to X_S \qquad (l = 0, 1, 2, \dots)$$

hold for all $\bar{q}^l \in X_R^{l+1}$. Let us denote:

$$\mathbf{A}_{q^j} \circ \cdots \circ \mathbf{A}_{q^0}(k_0, k_1, \dots) = (k_0^j, k_1^j, \dots) \equiv \bar{k}^j.$$

By lemma 8.2, the estimates:

$$d(\bar{x}^j(l), \bar{k}^j) \ < \ \delta$$

hold for $l = 1, 2$. So the estimates hold:

$$d(\bar{x}^j(1), \bar{x}^j(2)) \ \le d(\bar{x}^j(1), \bar{k}^j) + d(\bar{x}^j(2), \bar{k}^j) \ < \ 2\delta.$$

$\square$

Remark 8.2. In particular for two different stable extensions of the same automaton, we can still apply this and obtain uniform estimates between their orbits.

8.6 Uniform estimates for stable dynamics

Let:

$$\mathbf{A} : \begin{cases} \psi : Q \times S^{\alpha+1} \to S, \\ \phi : Q \times S^{\beta+1} \to Q \end{cases}$$

be an automaton, and choose a stable extension with the constants (δ, μ):

$$\begin{cases} \psi : \mathbb{R} \times \mathbb{R}^{\alpha+1} \to \mathbb{R}, \\ \phi : \mathbb{R} \times \mathbb{R}^{\beta+1} \to \mathbb{R} \end{cases}$$

represented by $(\max, +)$-functions. Let (ψ_t, ϕ_t) be the tropical correspondences to (ψ, ϕ), and M be the largest number of their components.

Let us consider the corresponding systems of the state dynamics:

$$\begin{cases} x(i,j+1) = \psi(\ y(i,j), x(i,j), \ldots, x(i+\alpha,j)), \\ y(i+1,j) = \phi(\ y(i,j), x(i,j), \ldots, x(i+\beta,j)), \end{cases}$$

$$\begin{cases} x'(i,j+1) = \psi_t(\ y'(i,j), x'(i,j), \ldots, x'(i+\alpha,j)), \\ y'(i+1,j) = \phi_t(\ y'(i,j), x'(i,j), \ldots, x'(i+\beta,j)) \end{cases}$$

with the same initial values:

$$x(i,0) = x'(i,0) = x_i, \quad y(0,j) = y'(0,j) = y^j$$

for $\bar{x} = \{x_i\}_i$ and $\bar{y} = \{y^j\}_j$ respectively.

Lemma 8.3. *Suppose $t_0 \gg 1$ satisfies the estimates:*

$$\mu\delta + 2\log_{t_0} M < \delta.$$

Then for all $t \geq t_0$ and any initial data with the uniform bounds:

$$d(\bar{x}, \bar{k}), \quad d(\bar{y}, \bar{q}) \;<\; \frac{\delta}{2}$$

for some $\bar{k} \in X_S$ and $\bar{q} \in X_Q$, their orbits satisfy the uniform estimates:

$$|x(i,j) - x'(i,j)|, \ |y(i,j) - y'(i,j)| < \frac{\delta}{2}.$$

Proof. Step 1: The first condition is satisfied for sufficiently large $t_0 \gg 1$, since $\mu < 1$ holds by stability assumption.

Let us denote the orbits by $(\{k_i^j\}, \{q_i^j\})$ determined by:

$$\begin{cases} k_i^{j+1} = \psi(q_i^j, k_i^j, \ldots, k_{i+\alpha}^j), \\ q_{i+1}^j = \phi(q_i^j, k_i^j, \ldots, k_{i+\beta}^j) \end{cases}$$

with the initial values $\bar{k}$ and $\bar{q}$ as above. By lemma 8.2, the uniform estimates hold for all i, j:

$$|x(i,j) - k_i^j|, \quad |y(i,j) - q_i^j| < \frac{\delta}{2}.$$

Let us consider $y'(i,0)$ and $x'(i,1)$. $y(0,0) = y'(0,0)$ holds, and let us verify the estimates:

$$|y'(i+1,0) - y(i+1,0)|, \quad |x(i,1) - x'(i,1)| \;<\; \frac{\delta}{2}$$

by induction on $i = 0, 1, 2, \ldots$ Suppose they hold up to i. Notice the estimates:

$$|y'(i,0) - q_i^0| \;\leq\; |y(i,0) - q_i^0| + |y'(i,0) - y(i,0)| \;<\; \delta.$$

So $y'(i,0)$ lies in the δ neighborhood of Q.

Then we have the estimates:

$$|y(i+1,0) - y'(i+1,0)|$$
$$= |\phi(y(i,0), x_i, \ldots, x_{i+\alpha}) - \phi_t(y'(i,0), x_i, \ldots, x_{i+\alpha})|$$
$$\leq |\phi(y(i,0), x_i, \ldots, x_{i+\alpha}) - \phi(y'(i,0), x_i, \ldots, x_{i+\alpha})|$$
$$+ |\phi(y'(i,0), x_i, \ldots, x_{i+\alpha}) - \phi_t(y'(i,0), x_i, \ldots, x_{i+\alpha})|$$
$$\leq \mu|y(i,0) - y'(i,0)| + \log_t M < \mu\frac{\delta}{2} + \log_t M < \frac{\delta}{2}.$$

where we used lemma 3.4 for the second inequality above. So we have verified the estimates $|y'(i,0) - y(i,0)| < \frac{\delta}{2}$ for all i by induction.

Next we have the estimates:

$$|x(i,1) - x'(i,1)| = |\psi(y(i,0), x_i, \ldots, x_{i+\alpha}) - \psi_t(y'(i,0), x_i, \ldots, x_{i+\alpha})|$$
$$\leq |\psi(y(i,0), x_i, \ldots, x_{i+\alpha}) - \psi(y'(i,0), x_i, \ldots, x_{i+\alpha})|$$
$$+ |\psi(y'(i,0), x_i, \ldots, x_{i+\alpha}) - \psi_t(y'(i,0), x_i, \ldots, x_{i+\alpha})|$$
$$\leq \mu|y(i,0) - y'(i,0)| + \log_t M < \mu\frac{\delta}{2} + \log_t M < \frac{\delta}{2}.$$

So we have verified the estimates $d(x'(i,1), x(i,1)) < \frac{\delta}{2}$.

Step 2: Let us put the sequences $\bar{x}^j \equiv (x(0,j), x(1,j), \ldots)$ and similarly for others. Let us verify the estimates:

$$d((\bar{y}')^j, \bar{y}^j), \quad d((\bar{x}')^{j+1}, \bar{x}^{j+1}) \quad < \quad \frac{\delta}{2}$$

by induction on $j = 0, 1, 2, \ldots$

We are done for $j = 0$ at step 1. Suppose the above estimates hold up to $j - 1$.

Let us start from the former case. $y(0,j) = y'(0,j)$ holds, and suppose the estimates $|y'(i,j) - y(i,j)| < \frac{\delta}{2}$ hold up to i. Then we have the estimates:

$$|y(i+1,j) - y'(i+1,j)| =$$
$$|\phi(y(i,j), x(i,j), \ldots, x(i+\beta, j)) - \phi_t(y'(i,j), x'(i,j), \ldots, x'(i+\beta, j))|$$
$$\leq |\phi(y(i,j), x(i,j), \ldots, x(i+\beta, j)) - \phi(y'(i,j), x'(i,j), \ldots, x'(i+\beta, j))|$$
$$+ |\phi(y'(i,j), x'(i,j), \ldots, x'(i+\beta, j)) - \phi_t(y'(i,j), x'(i,j), \ldots, x'(i+\beta, j))|$$
$$\leq \mu \max_{i \leq l \leq \beta+i} \{|y(i,j) - y'(i,j)|, |x(l,j) - x'(l,j)|\} + \log_t M$$
$$< \mu\frac{\delta}{2} + \log_t M < \frac{\delta}{2}.$$

So we have verified the estimates $d((\bar{y}')^j, \bar{y}^j) < \frac{\delta}{2}$ for all j by induction.

Next we have the estimates:

$|x(i, j+1) - x'(i, j+1)| =$
$|\psi(y(i,j), x(i,j), \ldots, x(i+\alpha, j)) - \psi_t(y'(i,j), x'(i,j), \ldots, x'(i+\alpha, j))|$
$\leq |\psi(y(i,j), x(i,j), \ldots, x(i+\alpha, j)) - \psi(y'(i,j), x'(i,j), \ldots, x'(i+\alpha, j))|$
$+ |\psi(y'(i,j), x'(i,j), \ldots, x'(i+\alpha, j)) - \psi_t(y'(i,j), x'(i,j), \ldots, x'(i+\alpha, j))|$
$\leq \mu \max_{i \leq l \leq \alpha+i} \{|y(i,j) - y'(i,j)|, |x(l,j) - x'(l,j)|\} + \log_t M$
$< \mu \frac{\delta}{2} + \log_t M < \frac{\delta}{2}.$

So we have verified the estimates $d((\bar{x}')^{j+1}, \bar{x}^{j+1}) < \frac{\delta}{2}$.

So we have completed the induction step on j. $\qquad\square$

8.7 Rational dynamics and change of automata

Let $S \subset \mathbb{R}$ be a finite set, and take $\bar{a} = (a_0, a_1, \ldots) \in X_S$. The *exponential sequence* is given by the sequence $t^{\bar{a}} = (t^{a_0}, t^{a_1}, \ldots)$ of positive numbers parametrized by $t > 1$. Its $C \geq 1$ neighborhood is given by the set:

$$N_C(t^{\bar{a}}) = \{\bar{z} \in X_{\mathbb{R}} : C^{-1} t^{a_i} < z_i < C t^{a_i}\} \supset t^{\bar{a}}.$$

For $l = 1, 2$, let:

$$\mathbf{A}_l : \begin{cases} \psi^l : Q \times S^{\alpha+1} \to S, \\ \phi^l : Q \times S^{\beta+1} \to Q \end{cases}$$

be two automata, and choose their stable extensions:

$$\begin{cases} \psi^l : \mathbb{R} \times \mathbb{R}^{\alpha+1} \to \mathbb{R}, \\ \phi^l : \mathbb{R} \times \mathbb{R}^{\beta+1} \to \mathbb{R} \end{cases}$$

by relative $(\max, +)$-functions with the constants (δ, μ).

Let (f_t^l, g_t^l) and (ψ_t^l, ϕ_t^l) be the tropical correspondences to (ψ^l, ϕ^l) respectively, and M be the largest number of the components.

Let us consider the state systems of the rational dynamics:

$$\begin{cases} z_i^{j+1}(l) = f_t^l(w_i^j(l), z_i^j(l), \ldots, z_{i+\alpha}^j(l)), \\ w_{i+1}^j(l) = g_t^l(w_i^j(l), z_i^j(l), \ldots, z_{i+\beta}^j(l)) \end{cases}$$

with the initial values $\bar{z}(l) = \{z_i(l)\}_i$ and $\bar{w}(l) = \{w^j(l)\}_j$ for $l = 1, 2$.

Theorem 8.1. *Suppose $\mathbf{A}_1$ and $\mathbf{A}_2$ are equivalent over some $R \subset Q$, and choose stable extensions with the constants (δ, μ).*

For any large $C \gg 1$, there exists $t_0 > 1$ so that for all $t \geq t_0$ and any initial value contained in:

$$\bar{z}(l) \in N_C(t^{\bar{k}}), \quad \bar{w}(l) \in N_C(t^{\bar{q}}) \qquad (l = 1, 2)$$

for some $\bar{k} \in X_S$ and $\bar{q} \in X_R$, then the uniform estimates hold:

$$\max\{\frac{z_i^j(1)}{z_i^j(2)}, \frac{z_i^j(2)}{z_i^j(1)}\} < C^4.$$

Let us compare this with proposition 6.3, where the constants grow double-exponentially with respect to i and j. In this case the orbit rates are uniformly bounded by a constant, which is much stronger since the Lipschitz constants are chosen to be contracting.

Proof. We split the proof into two steps.

Step 1: Notice that the pairs (ψ^l, ϕ^l) are (δ', μ)-stable for all $0 < \delta' \leq \delta$. Let us choose a large C so that the estimates hold:

$$M < C^{1-\mu}.$$

Then choose $t_0 \gg 1$ so that the estimates hold:

$$\log_{t_0} C \leq \frac{\delta}{2}.$$

Now let us choose and fix any $t \geq t_0$, and put $\delta' = \log_t C$. Then the estimates:

$$\mu\delta' + \log_t M = \mu \log_t C + \log_t M < \log_t C = \delta'$$

hold by the above inequality. So the condition in lemma 8.3 is satisfied. One may assume that the pairs (ψ^l, ϕ^l) are $(2\delta', \mu)$-stable.

Step 2: Let us put $x_i(l) = \log_t z_i(l)$ and $y^j(l) = \log_t w^j(l)$ for $l = 1, 2$, and consider the corresponding systems of the state dynamics:

$$\begin{cases} x_l(i, j+1) = \psi^l(y_l(i, j), x_l(i, j), \ldots, x_l(i+\alpha, j)), \\ y_l(i+1, j) = \phi^l(y_l(i, j), x_l(i, j), \ldots, x_l(i+\beta, j)), \end{cases}$$

$$\begin{cases} x'_l(i, j+1) = \psi^l_t(y'_l(i, j), x'_l(i, j), \ldots, x'_l(i+\alpha, j)), \\ y'_l(i+1, j) = \phi^l_t(y'_l(i, j), x'_l(i, j), \ldots, x'_l(i+\beta, j)) \end{cases}$$

with the initial values $\bar{x}(l) = \{x_i(l)\}_i$ and $\bar{y}(l) = \{y^j(l)\}_j$ respectively.

Then the estimates:

$$d(\bar{x}(l), \bar{k}) = \sup_i |x_i(l) - k_i| = \sup_i \log_t(\frac{z_i(l)}{t^{k_i}})^{\pm 1} < \log_t C = \delta'$$

hold. Similarly the estimates $d(\bar{y}(l), \bar{q}) < \delta'$ hold.

By corollary 8.1, the estimates:

$$|x_1(i, j) - x_2(i, j)| < 2\delta'$$

hold for all i, j.

On the other hand by lemma 8.3,

$$|x_l(i, j) - x'_l(i, j)| < \delta'$$

hold for $l = 1, 2$ and all i, j.

Then combining with these, we have the estimates:

$$|x'_1(i, j) - x'_2(i, j)| \le |x'_1(i, j) - x_1(i, j)| +$$
$$|x_1(i, j) - x_2(i, j)| + |x_2(i, j) - x'_2(i, j)|$$
$$< 4\delta' = \log_t C^4.$$

Since the equalities $|x'_1(i, j) - x'_2(i, j)| = \log_t(\frac{z_i^j(1)}{z_i^j(2)})^{\pm 1}$ hold, this verifies the desired estimates. $\square$

References

The method in this chapter is a more elaborate version to chapter 6, and it is applied to the Burnside group in chapter 9 [Kat9]. The idea of stability comes from the classical contraction principle in dynamical systems, see [Nit], [MS].

Chapter 9

Rational Burnside problem

Let us recall quasi-recursivity in 3.5. We have considered dynamics in the space time free case. More concretely let φ be a $(\max, +)$-function of n variable, and consider the one dimensional dynamics $x_N = \varphi(x_{N-n}, \ldots, x_{N-1})$. Let f_t be the tropical correspondence to φ, and consider the corresponding dynamics $z_N = f_t(z_{N-n}, \ldots, z_{N-1})$ with the initial data $x_0 = \log_t z_0$, $\ldots$, $x_{n-1} = \log_t z_{n-1}$.

We have verified that φ is always recursive over $\mathbb{R}$ whenever f_t is the case over $(0, \infty)$ for all $t > 1$. However the converse is not true in general. What we verified is that φ is recursive if and only if f_t is quasi-recursive.

Below we study recursivity in state dynamics.

9.1 Finite order group elements

Let:

$$\mathbf{A} : \begin{cases} \psi : Q \times S^{\alpha+1} \to S, \\ \phi : Q \times S^{\beta+1} \to Q \end{cases}$$

be an automaton, and

$$\begin{cases} \psi : \mathbb{R} \times \mathbb{R}^{\alpha+1} \to \mathbb{R}, \\ \phi : \mathbb{R} \times \mathbb{R}^{\beta+1} \to \mathbb{R} \end{cases}$$

be a stable extension with the constants (δ, μ).

Let (f_t, g_t) and (ψ_t, ϕ_t) be the corresponding functions to (ψ, ϕ) respectively, and M be the largest number of their components.

For $\bar{q}^m = (q^0, \ldots, q^m) \in X_Q^{m+1}$, let us denote l times iterations of $\bar{q}^m$ by $l\bar{q}^m \equiv (q^0, \ldots, q^m, q^0, \ldots, q^m, \ldots, q^0, \ldots, q^m) \in X_Q^{(m+1)l}$. We also denote the infinite times iterations of $\bar{q}^m$ by:

$$\bar{q}^m_{per} \equiv (q^0, \ldots, q^m, q^0, \ldots, q^m, \ldots, q^0, \ldots, q^m, \ldots) \in X_Q.$$

Let us consider the state dynamics:

$$\begin{cases} z_i^{j+1} = f_t(w_i^j, z_i^j, \dots, z_{i+\alpha}^j), \\ w_{i+1}^j = g_t(w_i^j, z_i^j, \dots, z_{i+\beta}^j). \end{cases}$$

Proposition 9.1. *Suppose* $\mathbf{A}_{\bar{q}^m} : X_S \to X_S$ *is of finite order with period* p:

$$\mathbf{A}_{p\bar{q}^m} = (\mathbf{A}_{\bar{q}^m}) \circ \cdots \circ (\mathbf{A}_{\bar{q}^m}) \equiv (\mathbf{A}_{\bar{q}^m})^p = \text{id}.$$

Then for any $C \geq 1$, *there exists* $t_0 > 1$ *so that for all* $t \geq t_0$ *and any initial value:*

$$\{z_i\}_i \subset N_C(t^S), \quad \{w^j\}_j \subset N_C(t^{\bar{q}_{per}^m})$$

the uniform bounds hold for all $i, j, l = 0, 1, 2, \dots$:

$$\left(\frac{z_i^j}{z_i^{j+p(m+1)l}}\right)^{\pm 1} \leq C^4$$

Proof. Let us choose large $t_0 > 1$ so that the estimates hold:

$$\log_{t_0} C, \ \mu\frac{\delta}{2} + \log_{t_0} M < \frac{\delta}{2}.$$

Recall that the pair (ψ, ϕ) is (δ', μ)-stable for any $0 < \delta' \leq \delta$. Let us fix $t \geq t_0$. Then by replacing δ by $\delta' = 2\log_t C$, one may assume the equality $\delta = 2\log_t C$.

By assumption, there is $\bar{k} = (k_0, k_1, \dots) \in X_S$ so that the initial value is contained as $\{z_i\}_i \in N_C(t^{\bar{k}})$.

Let us rewrite $\bar{q}_{per}^m = (q^0, q^1, \dots)$, and consider three state systems of dynamics:

$$\begin{cases} x(i, j+1) = \psi(y(i,j), x(i,j), \dots, x(i+\alpha, j)), \\ y(i+1, j) = \phi(y(i,j), x(i,j), \dots, x(i+\beta, j)), \end{cases}$$

$$\begin{cases} x'(i, j+1) = \psi_t(y'(i,j), x'(i,j), \dots, x'(i+\alpha, j)), \\ y'(i+1, j) = \phi_t(y'(i,j), x'(i,j), \dots, x'(i+\beta, j)), \end{cases}$$

$$\begin{cases} k_i^{j+1} = \psi(q_i^j, k_i^j, \dots, k_{i+\alpha}^j), \\ q_{i+1}^j = \phi(q_i^j, k_i^j, \dots, k_{i+\beta}^j) \end{cases}$$

with the initial values $x(i,0) = x'(i,0) = \log_t z_i$, $y(0,j) = y'(0,j) = \log_t w^j$, and $k_i^0 = k_i$, $q_0^j = q^j$. By the condition,

(1) the estimates $|x(i,0) - k_i|, |y(0,j) - q^j| < \frac{\delta}{2}$ hold, and

(2) periodicity $k_i^j = k_i^{j+p(m+1)l}$ holds for all i, j, l.

By lemma 8.2, the estimates $|x(i,j) - k_i^j| < \frac{\delta}{2}$ hold. By lemma 8.3, the estimates $|x(i,j) - x'(i,j)| < \frac{\delta}{2}$ hold. Combining these, we obtain the estimates:

$$|x'(i,j) - k_i^j| < \delta.$$

Then we have the estimates:

$$|x'(i,j) - x'(i, j + p(m+1)l)|$$
$$\leq |x'(i,j) - k_i^j| + |x'(i, j + p(m+1)l) - k_i^{j+p(m+1)}| < 2\delta = \log_t C^4.$$

Since the left hand side is equal to $\log_t(\frac{z_i^j}{z_i^{j+p(m+1)l}})^{\pm 1}$, the conclusion holds.

$\square$

9.2 Rational Burnside problem

The *Burnside problem* questions the existence of finitely generated and infinite torsion groups. A solution to the class of automata groups was constructed in chapter 7. Let us study how such a group property is reflected in the rational dynamics which corresponds to the Mealy automaton.

Let (f,g) be a pair of rational functions, and let us restate the state dynamics so that it is compatible with the expression of the dynamics by automata. For $w \in \mathbb{R}$, let:

$$\bar{\mathbf{A}}_w : \mathbb{R}^{\mathbb{N}} \to \mathbb{R}^{\mathbb{N}},$$
$$(z_0, z_1, \dots) \to (z_0', z_1', \dots)$$

be given by the rule:

$$\begin{cases} z_i' = f(w_i, z_i, \dots, z_{i+\alpha}), \\ w_{i+1} = g(w_i, z_i, \dots, z_{i+\beta}) \end{cases}$$

where $w_0 = w$. More generally for finite sequences $\bar{w}^l = (w^0, \dots, w^l) \in \mathbb{R}^{l+1}$, we denote:

$$\bar{\mathbf{A}}_{\bar{w}^l} \equiv \bar{\mathbf{A}}_{w^l} \circ \cdots \circ \bar{\mathbf{A}}_{w^0} : \mathbb{R}^{\mathbb{N}} \to \mathbb{R}^{\mathbb{N}}$$
$$\bar{\mathbf{A}}_{\bar{w}^l}(\bar{z}) = (z_0^l, z_1^l, \dots).$$

This is exactly the same as the state systems of the rational dynamics in 6.7 so that z_i^j above coincides with the ones in definition 6.5.

Let $X, Y \subset \mathbb{R}$ be two subsets, and denote $X^{\mathbb{N}} = \{(x_0, x_1, \dots); x_i \in X\}$. The state system is said to be *recursive* over (X, Y), if for any finite sequence:

$$\bar{w} = (w^0, \dots, w^l) \in Y^{l+1} \subset \mathbb{R}^{l+1}$$

there exist some $p = p(\bar{w}) \in \mathbb{N}$ so that the rational dynamics induce:

$$\bar{\mathbf{A}}_{\bar{w}} : X^{\mathbb{N}} \to X^{\mathbb{N}}$$

which are periodic $(\bar{\mathbf{A}}_{\bar{w}})^p = $ id of period p.

The *rational Burnside problem* questions the existence of pairs of rational functions whose state dynamics are recursive over some (X, Y), and $\bar{\mathbf{A}}_{\bar{w}}$ have mutually different periods with respect to infinitely many words $\bar{w} \in Y^*$.

Question 9.1. *Does such a pair of rational functions exist?*

So far the answer is not known.

Let us introduce a variant of the rational Burnside problem, where we need to use a pair of parametrized rational functions.

Firstly we introduce some notations. Let us consider two dynamics:

$$\bar{\mathbf{A}}_1, \bar{\mathbf{A}}_2 : X^{\mathbb{N}} \to X^{\mathbb{N}}$$

and denote their orbits by:

$$(\bar{\mathbf{A}}_l)^j(\bar{z}) = (z_0^j(l), z_1^j(l), \dots)$$

for $l = 1, 2$. Let us say that:

(1) $\bar{\mathbf{A}}_1$ and $\bar{\mathbf{A}}_2$ are *C-close* over $X \subset \mathbb{R}$, if for any $\bar{z} \in X^{\mathbb{N}} \subset \mathbb{R}^{\mathbb{N}}$, the corresponding orbits satisfy the uniform estimates:

$$\max \left(\frac{z_i^1(1)}{z_i^1(2)}, \frac{z_i^1(2)}{z_i^1(1)} \right) \leq C$$

for all $i = 0, 1, 2, \dots$.

(2) $\bar{\mathbf{A}}_1$ and $\bar{\mathbf{A}}_2$ are *C'-separated* over X, if there is some $\bar{z} \in X^{\mathbb{N}}$ so that the bounds hold for some i:

$$\max \left(\frac{z_i^1(1)}{z_i^1(2)}, \frac{z_i^1(2)}{z_i^1(1)} \right) \geq C'.$$

Let (f_t, g_t) be a pair of elementary rational functions, and consider the state system of the rational dynamics.

Definition 9.1. The state system of the parametrized rational dynamics by (f_t, g_t) is infinitely quasi-recursive if there exist a constant $C > 0$ and subsets:
$$X_t, \ Y_t \ \subset \ \mathbb{R}$$
so that for any $C' \geq 1$, the following two conditions hold for all $t \geq t_0 = t_0(C')$:

(1) For any word $\bar{w} \in Y_t^{l+1}$, there exist some $p = p(\bar{w}) \in \mathbb{N}$ such that $(\bar{\mathbf{A}}_{\bar{w}})^p$ and the identity are C-close over X_t.

(2) There are infinitely many words $\bar{w}^{(j)} \in Y_t^*$ such that $\bar{\mathbf{A}}_{\bar{w}^{(j)}}$ and $\bar{\mathbf{A}}_{\bar{w}^{(j')}}$ are mutually C'-separated over X_t for any $j \neq j'$.

Let $S = \{s_0, s_1\} \subset \mathbb{Z}$ be any embedding, and denote:
$$N_C(t^S) = \{z \in (0, \infty) : C^{-1} t^s \leq z \leq C t^s : s \in S\}.$$

Now we verify the existence of infinitely quasi-recursive dynamics:

Theorem 9.1. *There exists a pair of elementary rational functions (f_t, g_t) so that for any $C > 0$, the state dynamics is infinitely quasi-recursive over:*
$$(X_t, Y_t) = (N_C(t^S), N_C(t^Q)).$$

Proof. Let us choose another embedding $Q = \{q^0, \ldots, q^4\} \subset \mathbb{Z}$, and introduce the almost periodic exponential sequences:
$$N_C^{per}(t^Q) = \{(w^0, w^1, \ldots) : C^{-1} t^{q^j} \leq w^j \leq C t^{q^j}, \bar{q} = \bar{q}_{per}^m \in X_Q \text{ are periodic}\}.$$
which is a subspace of $N_C(t^Q)$.

Notice that if two points satisfy the inequalities $C^{-1} t^s \leq z, z' \leq C t^s$, then their ratios satisfy the bound $(\frac{z}{z'})^{\pm 1} \leq C^2$.

On the other hand if $C^{-1} t^{s_0} \leq z \leq C t^{s_0}$ and $C^{-1} t^{s_1} \leq z' \leq C t^{s_1}$ hold, then their ratio is separated by:
$$\left(\frac{z}{z'}\right)^{\pm 1} \geq C^{-2} t.$$

In particular if $t >> 1$ is sufficiently large, then $C^{-2} t \geq C'$ holds.

Let us consider a Mealy automaton $\mathbf{A}$ which produces the infinite torsion groups in theorem 7.1. By proposition 8.1, there exists a stable extension of $\mathbf{A}$. Let (f_t, g_t) be the pair of the corresponding elementary rational functions.

Because the group generated by $\{q^0, q^1\}$ is infinite torsion, the conclusion follows from theorem 8.1.

$\square$

Chapter 10

KdV equation and box-ball systems

The KdV (Korteweg–de Vries) equation is a main subject in mathematical physics, and there have been many researches on the analysis of their solutions. On the other hand the lamplighter group is also an important subject in group theory, and there have been many researches on this group particularly on random walks. Despite their importance, these subjects are rather separated from each other in the mathematical fields. So who could imagine that both the KdV equation and the lamplighter group can share structural similarities! Direct comparison of their structures would cause quite complicated situations, however we might be able to compare their frameworks after scale transform in tropical geometry, which makes their structures much simpler.

10.1 Discrete KdV equation

The KdV equation

$$\frac{\partial u}{\partial s}(x, s) + 6u \frac{\partial u}{\partial x}(x, s) + \frac{\partial^3 u}{\partial x^3}(x, s) = 0$$

plays an important role in integrable systems. It has particular solutions called 'soliton solutions', which exhibit interactions among multiple solitons.

Remark 10.1. Soliton solutions induce the Hirota's τ-function as

$$u = 2 \frac{\partial^2}{\partial x^2} \log \tau$$

which enables us to derive the famous bilinear equation of the KdV equation,

$$(D_s D_x + D_x^4)\tau \cdot \tau = 0,$$

which can be rewritten as

$$((\partial_s - \partial_{s'})(\partial_x - \partial_{x'}) + (\partial_x - \partial_{x'})^4)\tau(x,s)\tau(x',s')\big|_{x'=x,s'=s} = 0.$$

The KdV equation is incorporated in a huge structure governed by the τ-function, but it is beyond the scope of this book.

Let us introduce a discrete analog of the KdV equation.

Definition 10.1. A discrete KdV equation is given by:

$$\frac{1}{z_n^{t+1}} - \frac{1}{z_n^t} = z_{n+1}^{t+1} - z_{n-1}^t.$$

Proposition 10.1. *A continuous limit of the discrete KdV equation induces the KdV equation:*

$$u_s - \frac{1}{p^3}uu_x - \frac{1}{6p^6}u_{3x}.$$

Proof. Let us rewrite the equation as:

$$z_{n+1}^{t+\frac{1}{2}} - z_{n-1}^{t-\frac{1}{2}} = \frac{1}{z_n^{t+\frac{1}{2}}} - \frac{1}{z_n^{t-\frac{1}{2}}}$$

and rescale the parameters as:

$$n = \frac{s}{\epsilon^3}, \quad t = \frac{x}{\epsilon} - \frac{cs}{\epsilon^3},$$
$$z_n^t = p + \epsilon^2 u(x,s)$$

where c and p are constants satisfying $1 - 2c = 1/p^2$.

For example $n+1$ corresponds to $s + \epsilon^3$ and hence the pair $(n+1, t+\frac{1}{2})$ corresponds to $(s + \epsilon^3, x')$ where the equality holds:

$$\frac{x}{\epsilon} - \frac{cs}{\epsilon^3} + \frac{1}{2} = \frac{x'}{\epsilon} - \frac{c(s + \epsilon^3)}{\epsilon^3}.$$

Then we obtain a formula $x' = x + \epsilon(\frac{1}{2} + c)$.

By applying these change of variables into the defining equation above, one obtains the equation:

$$\epsilon^2 u\left(x + \frac{\epsilon}{2} + c\epsilon, s + \epsilon^3\right) - \epsilon^2 u\left(x - \frac{\epsilon}{2} - c\epsilon, s - \epsilon^3\right)$$
$$= \frac{1}{p + \epsilon^2 u(x + \frac{\epsilon}{2}, s)} - \frac{1}{p + \epsilon^2 u(x - \frac{\epsilon}{2}, s)}.$$

The Taylor expansion of the left hand side gives the formula at $\epsilon = 0$:

$$\text{l.h.s.} = \epsilon^2 \left\{ \epsilon(1 + c)u_x + 2\epsilon^3 u_s + \frac{\epsilon^3}{3}(1 + c)^3 u_{3x} \right\} + O(\epsilon^6).$$

The right hand side is by:

$$\text{r.h.s.} = \frac{1}{p + \epsilon^2 u + \frac{\epsilon^3}{2}u_x + O(\epsilon^4)} - \frac{1}{p + \epsilon^2 u - \frac{\epsilon^3}{2}u_x + O(\epsilon^4)}$$

$$= -\frac{\epsilon^3 u_x}{p^2}\left\{ 1 + \frac{2}{p}\epsilon^2 u + O(\epsilon^4)\right\}^{-1}$$

$$= -\frac{\epsilon^3 u_x}{p^2}\left\{ 1 - \frac{2}{p}\epsilon^2 u\right\} + O(\epsilon^7).$$

So if $p^2(1 + c) + 1 = 0$ holds, then we obtain the following formula:

$$2\epsilon^5\left(u_s - \frac{1}{p^3}uu_x - \frac{1}{6p^6}u_{3x}\right) + O(\epsilon^7) = 0.$$

Combining with our construction of contracting maps, we obtain a dynamical expansion from the cell automaton above to the KdV equation.

□

There is a structural similarity between the KdV equation and the Lotka–Volterra equation (see references below). Recall the Lotka–Volterra cell automaton in 6.2:

$$x_n^{t+1} = x_n^t + \max(0, x_{n+1}^t) - \max(0, x_{n-1}^{t+1}).$$

Its tropical correspondence is given by:

$$z_n^{t+1} = z_n^t \frac{1 + z_{n+1}^t}{1 + z_{n-1}^{t+1}}.$$

Lemma 10.1. *A continuous limit of the discrete Lotka–Volterra equation induces the Lotka–Volterra equation:*

$$u_n' = u_n(u_{n+1} - u_{n-1}).$$

Proof. Let us rescale the parameters as $z_n^t = \epsilon u_n(s)$ and $t = \frac{s}{\epsilon}$. Then insert the Taylor expansions into the defining equation:

$$\epsilon u_n + \epsilon^2 u_n' + O(\epsilon^3) = \epsilon u_n \frac{1 + \epsilon u_{n+1}}{1 + \epsilon u_{n-1} + \epsilon^2 u_{n-1}' + O(\epsilon^3)}.$$

We obtain the formula:

$$\epsilon(u_n' - u_n(u_{n+1} - u_{n-1})) + O(\epsilon^2) = 0.$$

□

The *modified KdV equation* (m-KdV) is given by the formula:

$$\frac{\partial v}{\partial s}(x, s) + 6v^2\frac{\partial v}{\partial x}(x, s) + \frac{1}{4}\frac{\partial^3 v}{\partial x^3}(x, s) = 0.$$

We also obtain it from a *discrete m-KdV equation*

$$w_{n+1}^{(t+1)}\frac{(1 + \gamma)w_n^{(t+1)} + \delta}{(1 + \delta)w_n^{(t+1)} + \gamma} = w_n^{(t)}\frac{(1 + \gamma)w_{n+1}^{(t)} + \delta}{(1 + \delta)w_{n+1}^{(t)} + \gamma},$$

under some suitable continuous limit.

10.2   From discrete KdV to BBS

Let us recall the discrete KdV equation $\dfrac{1}{u_{n+1}^{(t+1)}} - \dfrac{1}{u_n^{(t)}} = \delta\left(u_{n+1}^{(t)} - u_n^{(t+1)}\right)$,

and rewrite it as:

$$\frac{u_{n+1}^{(t)}}{v_{n+1}^{(t)}} \equiv \frac{1 - \delta u_{n+1}^{(t)} u_{n+1}^{(t+1)}}{u_{n+1}^{(t+1)}} = \frac{1 - \delta u_n^{(t)} u_n^{(t+1)}}{u_n^{(t)}} = \frac{u_n^{(t+1)}}{v_n^{(t)}}.$$

Then we obtain its subtraction free form:

$$\begin{cases} u_n^{(t)} u_n^{(t+1)} = (1/v_n^{(t)} + \delta)^{-1}, \\ v_{n+1}^{(t)} u_n^{(t+1)} = v_n^{(t)} u_{n+1}^{(t)}. \end{cases}$$

Under the boundary condition $u_{-n}^{(t)} = 1$ for all sufficiently large $n \gg 1$, we obtain the equality:

$$v_n^{(t)} = \frac{u_n^{(t)}}{u_{n-1}^{(t+1)}} v_{n-1}^{(t)} = \cdots = \prod_{j=-\infty}^{n-1} \frac{u_{j+1}^{(t)}}{u_j^{(t+1)}}.$$

Suppose that $u_n^{(t)}, \delta$ take positive values, then we introduce the following transformation of dependent variables:

$$u_n^{(t)} = \exp(B_n^{(t)}/\varepsilon), \qquad \delta = \exp(-1/\varepsilon).$$

By use of ultradiscrete limit or tropical transform, the subtraction free form of the discrete KdV equation induces the piecewise-linear system,

$$B_n^{(t+1)} = \min\left(1 - B_n^{(t)}, \sum_{j=-\infty}^{n-1} (B_j^{(t)} - B_j^{(t+1)})\right).$$

Lemma 10.2. *The above system is closed under the restriction of values in $\{0,1\}$ under the boundary condition $B_{-n}^{(t)} = 0$ for all large $n \gg 1$.*

Definition 10.2. Box-ball system (BBS) is an evolutional dynamics given by the above rule with the above boundary condition whose values take in $\{0,1\}$.

BBS is one of the ultradiscrete integrable systems. There is a geometric interpretation of BBS as follows: it is composed of an array of infinitely many boxes, finite number of balls in the boxes and a carrier of balls. Each box can contain at most one ball, and the carrier can transport an arbitrary number of balls. The evolution rule from time j to time $j + 1$ is described as follows. The carrier moves from left to right and passes each box. When

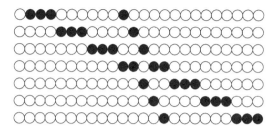

Fig. 10.1   A two-soliton interaction of the BBS

the carrier passes a box with a ball, then it gets the ball. When the carrier passes an empty box and if the carrier holds balls, it puts one ball into the box.

Let us describe BBS with carrier capacity k, which we denote by BBS$_k$. In this case the carrier can hold at most k balls. The only difference from the previous situation is that when the carrier already holds k balls and passes a box with a ball, then it does nothing.

Similar to the hierarchy of scale transform in the KdV–dKdV–BBS case, BBS with carrier capacity k can be obtained from the discrete modified KdV equation. Ultradiscrete limit or tropical transform as above induces the following piecewise-linear system:

$$\widetilde{B}_n^{(t+1)} = \min\left(1 - \widetilde{B}_n^{(t)}, \sum_{j=-\infty}^{n-1} (\widetilde{B}_j^{(t)} - \widetilde{B}_j^{(t+1)})\right)$$

$$+ \max\left(0, \sum_{j=-\infty}^{n} (\widetilde{B}_j^{(t)} - \widetilde{B}_{j-1}^{(t+1)}) - k\right).$$

This is also closed under the restriction of values in $\{0, 1\}$ under the boundary condition $\widetilde{B}_{-n}^{(t)} = 0$ for all large $n \gg 1$.

Box-ball system with carrier capacity k (BBS$_k$) is an evolutional dynamics given by the above rule with the above boundary condition whose values take in $\{0, 1\}$.

10.3   Soliton solutions

Let us describe how soliton solutions behave under scale transform. Let p_i and η_i^0 be constants, and denote:

$$\tau = 1 + \exp(\eta_1(x, s)) + \exp(\eta_2(x, s)) + c_{12} \exp(\eta_1(x, s) + \eta_2(x, s))$$

where $\eta_i(x, s) = p_i x - p_i^3 s + \eta_i^0$ with constants η_i^0 and $c_{12} = (\frac{p_i - p_j}{p_i + p_j})^2$.
τ is called a 2-soliton solution. In fact:

$$u = 2(\log \tau)_{xx}$$

satisfies the KdV equation.

Lemma 10.3. *Let $\mu(x) > 0$ be a differentiable and positive function. Then the formula holds:*

$$\frac{d^2}{dx^2} \log \mu = \lim_{\epsilon \to 0} \left\{ \frac{\mu(x - \epsilon)\mu(x + \epsilon)}{\mu(x)^2} - 1 \right\} / \epsilon^2.$$

Proof. Notice the formula $\frac{d^2}{dx^2} \log \mu = \frac{\mu_{xx}\mu - \mu_x^2}{\mu^2}$.

Let us take a Taylor expansion $\mu(x + \epsilon) = \mu(x) + \epsilon\mu_x(x) + \frac{\epsilon^2}{2}\mu_{xx}(x) + O(\epsilon^3)$, and insert as:

$$\frac{\mu(x - \epsilon)\mu(x + \epsilon)}{\mu(x)^2}$$

$$= \frac{(\mu(x) + \epsilon\mu_x(x) + \frac{\epsilon^2}{2}\mu_{xx}(x) + O(\epsilon^3))(\mu(x) - \epsilon\mu_x(x) + \frac{\epsilon^2}{2}\mu_{xx}(x) + O(\epsilon^3))}{\mu^2}$$

$$= 1 + \epsilon^2 \frac{\mu_{xx}\mu - \mu_x^2}{\mu^2} + O(\epsilon^3).$$

$\square$

It follows from lemma 10.3 that one can choose $\frac{\mu_{n-1}^t \mu_{n+1}^{t+1}}{\mu_n^t \mu_n^{t+1}}$ as a discrete analog of $\frac{d^2}{dx^2} \log \mu$.

Let us recall the discrete KdV equation $\dfrac{1}{u_{n+1}^{(t+1)}} - \dfrac{1}{u_n^{(t)}} = \delta\left(u_{n+1}^{(t)} - u_n^{(t+1)}\right)$, and denote:

$$\tau_n^t = 1 + h_1(n, t) + h_2(n, t) + c_{12}h_1(n, t)h_2(n, t)$$

where:

$$h_i(n, t) = (p_i)^n (q_i)^t h_i^0, \quad q_i = \frac{1 + \delta p_i^{-1}}{1 + \delta p_i}, \quad c_{12} = (\frac{p_1 - p_2}{p_1 p_2 - 1})^2$$

with constants h_i^0. One can check that

$$u_n^{(t)} \equiv \frac{\tau_{n-1}^t \tau_{n+1}^{t+1}}{\tau_n^t \tau_n^{t+1}}$$

satisfies the discrete KdV equation above.

It turns out that one can obtain soliton solutions to the Lotka–Volterra cell automaton in 6.2.1 from the tropical correspondence to τ_n^t above, which comes from the systematic study on Hirota's τ-function (see remark 10.1).

10.4 Diagrams of Mealy automata

Recall the Mealy automata diagram in chapter 7. We have also described the action induced from the diagram in Fig. 7.1.

Lemma 10.4. *The diagram expression of the BBS with carrier capacity k is given in Fig. 10.2.*

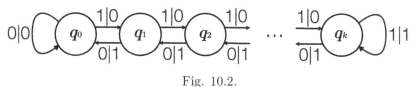

Fig. 10.2.

The (simple) BBS is obtained as the limiting case of the above automaton with $k \to \infty$.

The state q_j corresponds to the carrier with j balls. The action $\mathbf{A}_{q_j}$ represents time-evolution of BBS_k dynamics with the carrier with j balls as an initial state.

Proof. The state q_i corresponds to the situation when the carrier holds i balls. Thus we start at the state q_0. If we have 1 as the input we go from the state q_i to q_{i+1} if $i < k$ and we change 1 to 0. This corresponds to the fact that the carrier picks up the ball if the number of balls it already holds is $i < k$. If we have 0 as the input we go from the state q_i to q_{i-1} if $i > 0$ and change 1 to 0. This corresponds to the fact that the carrier releases the ball if the number of balls it already holds is at least 1. It remains to check the situation for q_0 with the input 0 and for q_k with the input 1. The first one corresponds to the carrier with 0 balls passing an empty box (it does nothing and still holds no balls) and the last one to the carrier with k balls passing a box with a ball (it does nothing and still holds k balls). $\square$

10.4.1 *BBS with carrier capacity $k = 1$*

We call $\mathrm{BBS}_{k=1}$ as *BBS translation*, since its dynamics behave in translation invariant way. The BBS translation diagram is represented in Fig. 10.3.

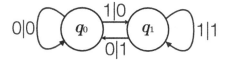

Fig. 10.3.

Recall that each state gives an action on the regular rooted tree T_2 in 6.1. Let $l^2(T_2)$ be the set of l^2 functions on the set of vertices of T_2.

Let a_i be the operators acting on $l^2(T_2)$ induced by $\mathbf{A}_{q_i}$, which are given by permutations of the coordinates for $i = 0, 1$. Let us restrict the actions of a_0 and a_1 on level sets $T_2^{(n)}$ of length n, in other words on the binary sequences of length n. They are represented by the $2^n \times 2^n$ matrices $a_0^{(n)}$ and $a_1^{(n)}$ respectively. They satisfy the following recurrence relations:

$$a_0^{(0)} = a_1^{(0)} = 1,$$

$$a_0^{(n+1)} = \begin{pmatrix} a_0^{(n)} & a_1^{(n)} \\ 0 & 0 \end{pmatrix}, \quad a_1^{(n+1)} = \begin{pmatrix} 0 & 0 \\ a_0^{(n)} & a_1^{(n)} \end{pmatrix}.$$

Definition 10.3. The approximation of the transition operator is given by a family of symmetric matrices:

$$M_{k=1}^{(n)} = \frac{1}{4} \left(a_0^{(n)} + a_0^{(n)*} + a_1^{(n)} + a_1^{(n)*} \right).$$

10.4.2   *BBS with carrier capacity $k = 2$*

Let us consider the $k = 2$ case.

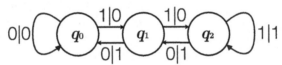

Fig. 10.4.

Let a_i be the operators acting on $l^2(T_2)$ induced by $\mathbf{A}_{q_i}$ for $i = 0, 1, 2$. The restrictions of these actions on level sets $T_2^{(n)}$ of length n are also represented by the $2^n \times 2^n$ matrices $a_0^{(n)}, a_1^{(n)}$ and $a_2^{(n)}$, which satisfy the following recurrence relations:

$$a_0^{(0)} = a_1^{(0)} = a_2^{(0)} = 1,$$

$$a_0^{(n+1)} = \begin{pmatrix} a_0^{(n)} & a_1^{(n)} \\ 0 & 0 \end{pmatrix}, \quad a_1^{(n+1)} = \begin{pmatrix} 0 & a_2^{(n)} \\ a_0^{(n)} & 0 \end{pmatrix}, \quad a_2^{(n+1)} = \begin{pmatrix} 0 & 0 \\ a_1^{(n)} & a_2^{(n)} \end{pmatrix}.$$

The *approximation of the transition operator* is given by a family of finite matrices:

$$M_{k=2}^{(n)} = \frac{1}{6} \left(a_0^{(n)} + a_0^{(n)*} + a_1^{(n)} + a_1^{(n)*} + a_2^{(n)} + a_2^{(n)*} \right).$$

Later we will verify that these operators satisfy stochastic property, and analyze their spectral properties in detail for $k = 1$ theoretically and $k \geq 2$ numerically.

10.5 Scale transform around KdV

There is a structural similarity between the KdV and Lotka–Volterra equations. The Lotka–Volterra equation is given by:

$$\frac{d}{dt} v_n = v_n (v_{n+1} - v_{n-1}).$$

The modified KdV equation is given by the equation $v_s = v^2 v_x + v_{3x}$. If we replace v by $\tilde{v} = -1 + v$, then $\tilde{v}$ satisfies the equation $\tilde{v}_s = (1 + \tilde{v})^2 \tilde{v}_x + \tilde{v}_{3x}$. It has the form $u_s = (\alpha u + \beta u^2) u_x + u_{3x}$ if we introduce another change of variable from s to $s + x$. Then we obtain the KdV equation by letting $\beta \to 0$. Its discrete version:

$$\frac{d}{ds} v_n = (\alpha v_n + \beta v_n^2)(v_{n+1} - v_{n-1})$$

is denoted as L-V + KdV lattice in Fig. 10.5.

The KdV lattice is a semi-discrete version of the KdV equation:

$$\frac{d}{ds} \left(\frac{1}{v_n} \right) = v_{n+1} - v_{n-1}$$

so that the space and time variables are discrete and continuous respectively. This corresponds to $\alpha = 0$ and $\beta = -1$ above, while the Lotka–Volterra equation is obtained by letting $\alpha = 1$ and $\beta = 0$.

One can obtain the d-KdV equation when the time variable is also discretized. As a supplement, let us induce the d-KdV from the KdV lattice. Consider a fully discretized version of the KdV lattice:

$$\frac{1}{w_n^{t+1}} - \frac{1}{w_n^t} = w_{n+1}^t - w_{n-1}^{t+1}.$$

A change of variables as $z_{n+t}^t = w_n^t$ replaces the equation by $\frac{1}{z_{t+1+n}^{t+1}} - \frac{1}{z_{t+n}^t} = z_{t+n+1}^t - z_{t+n}^{t+1}$, which can be rewritten as:

$$\frac{1}{z_{m+1}^{t+1}} - \frac{1}{z_m^t} = z_{m+1}^t - z_m^{t+1}.$$

In general the scale transform is uniquely determined from discrete to continuous, however the converse is not the case. For example consider the discrete KdV equation, which is derived from the KdV lattice here. One may wonder whether another equation such as $\frac{1}{w_n^{t+1}} - \frac{1}{w_n^t} = w_{n+1}^t - w_{n-1}^t$ can be used. Actually both the discrete KdV and this one converge to the same KdV lattice. There is no canonical way to discretize differential equations, and optimal ways depend on the structure of the equations in general. In any situation, discrete dynamics should inherit some kinds of characteristics from the original differential equations.

One way is to choose specific equations among various choices of discrete ones by focusing on some characteristics of differential equations. It is the case for integrable systems whose characteristics are their conserved quantities, and in the KdV case, one can say that the discrete KdV is the correct choice. This is the fundamental viewpoint in integrable systems.

Another way is to focus on quite rough analytic properties of solutions to differential equations so that any choice of discretization can be applied. Our scope in this book covers construction of class differential equations for more general cases including non integrable systems. We develop rough analysis of the solutions to different partial differential equations in later chapters.

Remark 10.2. Let us give a comment on the words 'tropical' and 'ultra-discrete'. Essentially both imply the same process of scale transform, but historically they were born in different mathematical fields. The former was introduced in real algebraic geometry and the latter in integrable systems concerning KdV and related matters. This book focuses on geometry and analysis of scale transform, and hence we will mostly use the word tropical geometry.

References

Miura transform builds a bridge between KdV and m-KdV equations. This is a version of discrete Miura transform. Figure 10.5 gives the diagram of the relation between various equations and transformations.

$$
\begin{array}{ccccc}
\textbf{KdV} & \leftarrow & \textbf{KdV lattice} & \xrightarrow{\text{[TH]}} & \textbf{BBS} \\
 & \nwarrow & \updownarrow \text{ d-Miura} & \nearrow \text{[TTMS]} & \\
\uparrow \text{Miura} & & \textbf{Lotka–Volterra} & & \uparrow \ k \to \infty \\
 & & & & \\
\textbf{m-KdV} & \leftarrow & \textbf{L-V + KdV lattice} & \xrightarrow{\text{[TM]}} & \textbf{BBS}_k
\end{array}
$$

Fig. 10.5.

Box-ball systems (BBS) were introduced in [TS]. The scaling limit from Lotka–Volterra to BBS was given in [TTMS], while the scaling limit from Lotka–Volterra to BBS_k was given in [TM]. Tsujimoto and Hirota [TH] obtained the scaling limit from the KdV lattice to the BBS passing through d-KdV. BBS diagrams have been presented in [KTZ].

Chapter 11

Spectral similarity between BBS and lamplighter group

11.1 Markov operator

Let Γ be a finitely generated group equipped with a generating set. An associated connected graph is called the Cayley graph. Its vertices are the elements of the group, and two vertices are connected by an edge if and only if an element in the generating set or its inverse gives a transition between these two elements.

Let $\deg(v) \in \mathbb{N}$ be the number of edges emanating from v, which is a non negative integer. The *transition matrix* is a matrix $M = (m_{u,v})_{u,v \in \Gamma}$ (of infinite size if Γ is infinite) given by:

$$m_{u,v} = \begin{cases} \deg(u)^{-1} & \text{if } u \text{ and } v \text{ are connected by an edge,} \\ 0 & \text{otherwise.} \end{cases}$$

This is symmetric and satisfies the stochastic property in the sense that the sum of each row and the sum of each column are both equal to one. So this gives a measure of the path space on Γ.

Simple random walk (SRW) on Γ is the Markov chain given by a sequence X_n of Γ-valued random variables with Markov property:

$$\Pr[X_n = v : X_k = u_k, 0 \le k \le n - 1] = \Pr[X_n = v : X_{n-1} = u_{n-1}]$$

for any $u_0, \ldots, u_{n-1} = u, v \in \Gamma$. So X_n is the random position at time n as we move along the edges with probability $m_{u,v}$. We will also call M the *Markov operator*.

SRW has been studied extensively in geometry and analysis. In particular its spectral distribution is closely related to the analytic structure of Γ. For example it is known that Γ is amenable if and only if its spectral radius is 1. This motivates us to compute the eigenvalues of the transition operators, but in general it is not so easy.

The Markov operator is regarded as an element in the group ring $\mathbb{Q}\Gamma$. Let $\{q_1, \ldots, q_l\}$ be a generating set. Then one has the following expression:

$$M = \frac{1}{2l}(q_1 + q_1^{-1} + q_2 + q_2^{-1} + \cdots + q_l + q_l^{-1}).$$

If a group admits a faithful unitary representation, then it can be rewritten as:

$$M = \frac{1}{2l}(q_1 + q_1^* + q_2 + q_2^* + \cdots + q_l + q_l^*).$$

In the case of an automata group, there are two canonical unitary representations on $l^2(\Gamma)$ and $l^2(T_2)$. If the action on the tree is spherically transitive, then it turns out that the spectra of these coincide with each other [GZ].

In general a Mealy automaton gives a semi-group rather than a group. In the semi-group case, the second operator still exists. Let us interpret this operator: if one moves from u to v in probability $m_{u,v}$ in a semi-group, then let us allow the reverse move with the same probability. The corresponding transition matrix is given by the above formula.

Recall that an automata semi-group has a canonical generating set.

Definition 11.1. Let $\mathbf{A}$ be a Mealy automaton with the states $\{q_0, \ldots, q_l\}$ and the alphabets $\{s_0, \ldots, s_k\}$. Let Γ be the associated semi-group.

The transition operator is given by:

$$M = \frac{1}{2(l+1)}(q_0 + q_0^* + q_1 + q_1^* + \cdots + q_l + q_l^*)$$

which acts on $l^2(T_{k+1})$ as a self-adjoint bounded operator.

As we have already given for particular cases of $\text{BBS}_{k=1}$ and $\text{BBS}_{k=2}$, in general the restriction of the actions by the states on each level set of the tree gives a family of matrices:

$$M^{(n)} = \frac{1}{2(l+1)}(q_0^{(n)} + (q_0^{(n)})^* + q_1^{(n)} + (q_1^{(n)})^* + \cdots + q_l^{(n)} + (q_l^{(n)})^*)$$

We call this family as the *approximation of the transition operator*.

11.2 Markov operators for $\text{BBS}_{k=1}$ and lamplighter group

Let us state the main result in chapter 11 as follows:

Theorem 11.1. *(1) The spectra of the Markov operators coincide with each other between the lamplighter automata group and $\text{BBS}_{k=1}$. It is totally discrete and dense in $[-1, 1]$.*

(2) The approximations of the transition operators of BBS$_k$ satisfy stochastic property for all $k \geq 1$.

The second property has been known for the lamplighter automata group [GZ].

The Markov operator is a self-adjoint operator. Recall in linear algebra that a symmetric matrix can be diagonalizable. The above theorem might suggest both Markov operators for BBS$_{k=1}$ and lamplighter can be conjugate from each other by an invertible operator. Actually the answer is both yes and no.

Proposition 11.1. *(1) These operators cannot be conjugated by any tree automorphism.*

(2) There is a permutation which conjugate these operators.

This is rather satisfactory, since tree automorphism reflects dynamical properties and permutation automorphism is more algebraic. Dynamically BBS and the lamplighter group are very different, but their cell diagrams seem to resemble each other.

We will not give a proof for (2) (refer to [KTZ] for details).

Let us give a proof for (1). If there were an automorphism of the tree which would conjugate two operators on some level n it would also conjugate these operators on the previous levels. Thus it is enough to prove the statement for the level $n = 2$. For this level the operator corresponding to the BBS system has $(2, 0, 0, 2)$ on the diagonal and the operator corresponding to the lamplighter has $(0, 0, 2, 2)$ on the diagonal. The last one under the tree automorphism can be transformed to itself or $(2, 2, 0, 0)$ only.

11.3 Spectral computation for BBS translation ($k = 1$)

Stochastic property is closely related to random walk on each level set of the binary tree. On the other hand the structure of the random walk heavily depends on their spectral distribution (see [Woe]). First we compute the spectral distribution of the transition operator for BBS$_{k=1}$:

$$M_{k=1}^{(n)} = \frac{1}{4} \left(a_0^{(n)} + a_0^{(n)*} + a_1^{(n)} + a_1^{(n)*} \right).$$

Define the counting spectral measures of $M_k^{(n)}$, i.e. $\sigma_k^{(n)} : [0,1] \to [0,1]$ and for $x \in [0,1]$ by:

$$\sigma_k^{(n)}(x) = \frac{\sharp\left\{\lambda \in \mathrm{Sp}(M_k^{(n)}) \mid \lambda \leq 2(k+1)\cos(\pi x)\right\}}{\sharp\left\{\mathrm{Sp}(M_k^{(n)})\right\}}.$$

Let us denote the multiplicity of eigenvalue λ of $M_k^{(n)}$ by $m^{(n;k)}(\lambda)$.

We consider the case $k = 1$ and provide the computation of eigenvalues of $M_{k=1}^{(n)}$. Our computation on the spectra verify the following:

Theorem 11.2.

$$Sp\left(M_{k=1}^{(n)}\right) = Sp\left(\frac{1}{4}\sum_{j=0}^{1}\left(a_j^{(n)} + a_j^{(n)*}\right)\right)$$

$$= \left\{1 \cup \cos\left(\frac{p}{q}\pi\right)\middle|\, p,q \in \mathbb{N}, 1 \leq p < q \leq n+1\right\}.$$

If p and q are mutually prime, then the multiplicity of eigenvalue $\cos\left(pq^{-1}\pi\right)$, denoted by $m_{p,q}^{(n;1)}$, is given by

$$m_{p,q}^{(n;1)} = \left\lfloor 2^n\left(\frac{2^{-q} - 2^{-q\left(\left[\frac{n}{q}\right]+1\right)}}{1 - 2^{-q}}\right)\right\rfloor.$$

The proof consists of computating the characteristic polynomials.
In order to simplify the notation we let $a_n = a_0^{(n)}$ and $b_n = a_1^{(n)}$.

Lemma 11.1. *For every n*

$$a_n a_n^* + b_n b_n^* = 2Id_{2^n}.$$

Proof. We have

$$a_{n+1}a_{n+1}^* = \begin{pmatrix} a_n & b_n \\ 0 & 0 \end{pmatrix}\begin{pmatrix} a_n^* & 0 \\ b_n^* & 0 \end{pmatrix} = \begin{pmatrix} a_n a_n^* + b_n b_n^* & 0 \\ 0 & 0 \end{pmatrix}$$

$$b_{n+1}b_{n+1}^* = \begin{pmatrix} 0 & 0 \\ a_n & b_n \end{pmatrix}\begin{pmatrix} 0 & a_n^* \\ 0 & b_n^* \end{pmatrix} = \begin{pmatrix} 0 & 0 \\ 0 & a_n a_n^* + b_n b_n^* \end{pmatrix}$$

and the statement follows by induction. $\square$

Let us check a basic fact in linear algebra:

Assertion 11.1.

$$\det\begin{pmatrix} A & B \\ C & D \end{pmatrix} = \det(AD - CB)$$

provided that A commutes with C.

Proof. Let us verify the conclusion when A is invertible. Because the map:

$$\begin{pmatrix} A & B \\ C & D \end{pmatrix} \to \det\begin{pmatrix} A & B \\ C & D \end{pmatrix} - \det(AD - CB)$$

is continuous from the set of square matrices to $\mathbb{R}$, and because the set of invertible matrices is dense in all the square matrices, the general case also follows by continuity.

Then we have the equality:

$$\begin{pmatrix} A & B \\ C & D \end{pmatrix} = \begin{pmatrix} A & 0 \\ 0 & 1 \end{pmatrix}\begin{pmatrix} 1 & 0 \\ C & 1 \end{pmatrix}\begin{pmatrix} 1 & A^{-1}B \\ 0 & D - CA^{-1}B \end{pmatrix}.$$

Notice that we have not yet used commutativity. Then

$$\det\begin{pmatrix} A & B \\ C & D \end{pmatrix} = \det(A)\det(D - CA^{-1}B)$$

$$= \det(AD - ACA^{-1}B) = \det(AD - CB).$$

$\square$

Proof of Theorem 11.2.

Step 1: Let:

$$\Phi_n(\lambda, \mu) = \det(a_n + a_n^* + b_n + b_n^* - \frac{1}{2}\mu(a_n b_n^* + b_n a_n^*) - \lambda Id_{2^n}).$$

$\Phi_n(\lambda, 0)$ is equal to $\det(4M_n - \lambda)$, and hence solutions to the equation $\Phi_n(\lambda, 0) = 0$ concedes with the set of the eigenvalues of $4M_n$.

It follows from proposition 11.2 below that M_n satisfies the stochastic property, which implies that the sum of each row or each column is equal to 1. Hence the eigenvalues are in $[-4, 4]$. In particular they can be written as $4\cos z$ for some $z \in [0, \pi]$.

Step 2: Using lemma 11.1, we have the equalities:

$\Phi_{n+1}(\lambda, \mu)$

$$= \det(a_{n+1} + a_{n+1}^* + b_{n+1} + b_{n+1}^* - \frac{1}{2}\mu(a_{n+1}b_{n+1}^* + b_{n+1}a_{n+1}^*) - \lambda Id_{2^{n+1}})$$

$$= \det\begin{pmatrix} a_n + a_n^* - \lambda & b_n + a_n^* - \frac{1}{2}\mu(a_n a_n^* + b_n b_n^*) \\ a_n + b_n^* - \frac{1}{2}\mu(a_n a_n^* + b_n b_n^*) & b_n + b_n^* - \lambda \end{pmatrix}$$

$$= \det\begin{pmatrix} a_n + a_n^* - \lambda & b_n + a_n^* - \mu \\ a_n + b_n^* - \mu & b_n + b_n^* - \lambda \end{pmatrix}$$

$$= \det\begin{pmatrix} a_n - b_n - \lambda + \mu & b_n + a_n^* - \mu \\ a_n - b_n + \lambda - \mu & b_n + b_n^* - \lambda \end{pmatrix}$$

$$= \det\begin{pmatrix} 2\mu - 2\lambda & a_n^* - b_n^* - \mu + \lambda \\ a_n - b_n + \lambda - \mu & b_n + b_n^* - \lambda \end{pmatrix}.$$

Let us apply assertion 11.1 for the above matrix. In fact $2\mu - 2\lambda$ is scalar and hence it satisfies the assumption in assertion 11.1. Then we get the equalities:

$$
\begin{aligned}
\Phi_{n+1}(\lambda, \mu) &= \det((2\mu - 2\lambda)(b_n + b_n^* - \lambda) \\
&\quad - (a_n - b_n + \lambda - \mu)(a_n^* - b_n^* - \mu + \lambda)) \\
&= \det((\mu - \lambda)(a_n + a_n^* + b_n + b_n^*) \\
&\quad - \frac{1}{2}2(a_n b_n^* + b_n a_n^*) + (-2 + \lambda^2 - \mu^2)Id_{2^n}).
\end{aligned}
$$

Therefore:

$$
\Phi_{n+1}(\lambda, \mu) = (\mu - \lambda)^{2^n} \Phi_n \left(\frac{2 - \lambda^2 + \mu^2}{\mu - \lambda}, \frac{-2}{\mu - \lambda} \right)
$$

with $\Phi_0 = 4 - \lambda - \mu$.

This is exactly the formula from [GZ] which leads to the explicit computation of all eigenvalues.

Step 3: Let:

$$
\lambda' = -\frac{\lambda^2 - \mu^2 - 2}{\mu - \lambda}, \quad \mu' = -\frac{2}{\mu - \lambda}.
$$

They satisfy the equalities:

$$
\lambda' + \mu' = \lambda + \mu,
$$

$$
\mu' - \lambda' = -\mu - \lambda - \frac{4}{\mu - \lambda}.
$$

Let us consider a family $\{P_k(\lambda, \mu), Q_k(\lambda, \mu)\}_{k \geq 1}$ with $P_1 = \mu - \lambda$ and $Q_1 = 1$ such that their ratios satisfy the relation:

$$
\frac{P_{k+1}}{Q_{k+1}}(\lambda, \mu) = \frac{P_k(\lambda', \mu')}{Q_k(\lambda', \mu')}.
$$

We claim the equalities hold:

$$
\Phi_n(\lambda, \mu) = (4 - \lambda - \mu) \prod_{k=1}^{n} (\frac{P_k}{Q_k}(\lambda, \mu))^{2^{n-k}}.
$$

In fact

$$
\Phi_1 = (\mu - \lambda)\Phi_0 = (\mu - \lambda)(4 - \mu - \lambda) = (4 - \mu - \lambda)\frac{P_0}{Q_0}.
$$

Suppose the formulas hold up to n. Then we have:

$$\Phi_{n+1}(\lambda,\mu) = (\mu-\lambda)^{2^n}\Phi_n(\lambda',\mu')$$

$$= \left(\frac{P_1}{Q_1}(\lambda,\mu)\right)^{2^n}(4-\lambda-\mu)\prod_{k=1}^{n}\left(\frac{P_k}{Q_k}(\lambda',\mu')\right)^{2^{n-k}}$$

$$= \left(\frac{P_1}{Q_1}(\lambda,\mu)\right)^{2^n}(4-\lambda-\mu)\prod_{k=1}^{n}\left(\frac{P_{k+1}}{Q_{k+1}}(\lambda,\mu)\right)^{2^{n-k}}$$

$$= (4-\lambda-\mu)\prod_{k=1}^{n+1}\left(\frac{P_k}{Q_k}(\lambda,\mu)\right)^{2^{n+1-k}}$$

Step 4: We claim that their ratios satisfy another formula:

$$\frac{P_{k+1}}{Q_{k+1}}(\lambda,\mu) = -\mu-\lambda-\frac{4}{\frac{P_k}{Q_k}(\lambda,\mu)}.$$

Actually $\frac{P_1}{Q_1}(\lambda,\mu) = \mu-\lambda$, and:

$$\frac{P_2}{Q_2}(\lambda,\mu) = \frac{P_1}{Q_1}(\lambda',\mu') = -\mu-\lambda-\frac{4}{\mu-\lambda}.$$

Suppose the formulas hold up to $k-1$. Then we have:

$$\frac{P_{k+1}}{Q_{k+1}}(\lambda,\mu) = \frac{P_k}{Q_k}(\lambda',\mu') = -\mu'-\lambda'-\frac{4}{\frac{P_{k-1}}{Q_{k-1}}(\lambda',\mu')}$$

$$= -\mu-\lambda-\frac{4}{\frac{P_{k-1}}{Q_{k-1}}(\lambda',\mu')} = -\mu-\lambda-\frac{4}{\frac{P_k}{Q_k}(\lambda,\mu)}.$$

So it also holds for k, and we have verified the claim.

Step 5: Now let $\mu = 0$. Then we have the equality:

$$\frac{P_{k+1}}{Q_{k+1}}(\lambda,0) = -\lambda-\frac{4Q_k}{P_k}(\lambda,0) = \frac{-\lambda P_k - 4Q_k}{P_k}.$$

Later on we omit the variable $(\lambda,0)$, and just write P_k or Q_k.

Let us solve the matrix equations:

$$\begin{pmatrix} P_{k+1} \\ Q_{k+1} \end{pmatrix} = \begin{pmatrix} -\lambda & -4 \\ 1 & 0 \end{pmatrix}\begin{pmatrix} P_k \\ Q_k \end{pmatrix} = \begin{pmatrix} -\lambda & -4 \\ 1 & 0 \end{pmatrix}^2\begin{pmatrix} P_{k-1} \\ Q_{k-1} \end{pmatrix}$$

$$= \cdots = \begin{pmatrix} -\lambda & -4 \\ 1 & 0 \end{pmatrix}^k\begin{pmatrix} 1 \\ 0 \end{pmatrix}.$$

It is not so difficult to transform the above matrix to a diagonal matrix as:

$$
\begin{pmatrix} -\lambda & -4 \\ 1 & 0 \end{pmatrix}^k = \begin{pmatrix} \frac{8}{-\lambda+\sqrt{\lambda^2-16}} & \frac{-8}{\lambda+\sqrt{\lambda^2-16}} \\ 1 & 1 \end{pmatrix} \times \begin{pmatrix} \left(\frac{8}{-\lambda+\sqrt{\lambda^2-16}}\right)^k & 0 \\ 0 & \left(\frac{-8}{\lambda+\sqrt{\lambda^2-16}}\right)^k \end{pmatrix}
$$

$$
\times \begin{pmatrix} \frac{-1}{\sqrt{\lambda^2-16}} & \frac{1}{2}-\frac{\lambda}{2\sqrt{\lambda^2-16}} \\ \frac{1}{\sqrt{\lambda^2-16}} & \frac{1}{2}+\frac{\lambda}{2\sqrt{\lambda^2-16}} \end{pmatrix}
$$

Let $\lambda = 4\cos z$ as in step 1. Then the above matrix is equal to the following:

$$
\begin{pmatrix} \frac{2}{-\cos z+i\sin z} & \frac{-2}{\cos z+i\sin z} \\ 1 & 1 \end{pmatrix} \times \begin{pmatrix} \left(\frac{2}{-\cos z+i\sin z}\right)^k & 0 \\ 0 & \left(\frac{-2}{\cos z+i\sin z}\right)^k \end{pmatrix}
$$

$$
\times \begin{pmatrix} \frac{-1}{4i\sin z} & \frac{1}{2}-\frac{\cos z}{2i\sin z} \\ \frac{1}{4i\sin z} & \frac{1}{2}+\frac{\cos z}{2i\sin z} \end{pmatrix}.
$$

So we can compute P_k and Q_k as:

$$
P_k = \frac{2^k \sin z(k+1)}{\sin z}, \qquad Q_k = \frac{2^{k-1}\sin zk}{\sin z}.
$$

Then:

$$
\Phi_n(\lambda,0) = (4-4\cos z)\prod_{k=1}^{n}\left(\frac{P_k}{Q_k}\right)^{2^{n-k}} = (4-4\cos z)2^n \prod_{k=1}^{n}\left(\frac{2^k\sin z(k+1)}{\sin kz}\right)^{2^{n-k}}
$$

$$
= (4-4\cos z)2^n \left(\frac{1}{\sin z}\right)^{2^{n-1}}\left(\prod_{k=2}^{n}(\sin zk^{2^{n-k}})\right)\sin z(n+1).
$$

It takes zero exactly when $z = \frac{p}{q}\pi$ for $p,q \in \mathbb{N}$ with $1 \le p < q \le n+1$. It is not so difficult to check the formulas for their multiplicities.

This completes the proof of theorem 11.2.

11.4 Numerical computation of spectra for BBS ($k \ge 2$)

Computation of the characteristic polynomials of $M_k^{(n)}$ for $k \ge 2$ is not so easy. Actually the extra appearance of single roots seems to create continuous spectra as $n \to \infty$. On the other hand if we draw distributions of multiple eigenvalues as given in Tables 11.1 to 11.5, then we can find some structural similarity in their appearance. We will present some conjectures on their distributions after some observations.

Let us compare the histograms of the spectral distributions for $k = 1$ and 2. Figures 11.1 and 11.2 present the histograms of the distributions of

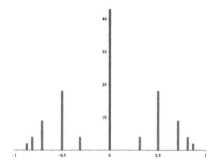 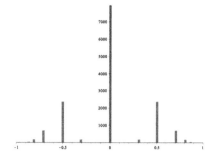

Fig. 11.1   Distribution of the multiple eigenvalues of $M_{k=1}^{(7)}$

Fig. 11.2   Distribution of the multiple eigenvalues of $M_{k=2}^{(14)}$

Table 11.1   Multiplicities of non-negative eigenvalues for $k = 1$

n	$m_{1,2}^{(n;1)}$	$m_{1,3}^{(n;1)}$	$m_{1,4}^{(n;1)}$	$m_{1,5}^{(n;1)}$	$m_{2,5}^{(n;1)}$	$m_{1,6}^{(n;1)}$	$m_{1,7}^{(n;1)}$	$m_{2,7}^{(n;1)}$	$m_{3,7}^{(n;1)}$	$m_{1,8}^{(n;1)}$	$\cdots$	$m_{5,11}^{(n;1)}$
1	1	0	0	0	0	0	0	0	0	0		0
2	1	1	0	0	0	0	0	0	0	0		0
3	3	1	1	0	0	0	0	0	0	0		0
4	5	2	1	1	1	0	0	0	0	0		0
5	11	5	2	1	1	1	0	0	0	0		0
6	21	9	4	2	2	1	1	1	1	0		0
7	43	18	9	4	4	2	1	1	1	1		0
8	85	37	17	8	8	4	2	2	2	1		0
9	171	73	34	17	17	8	4	4	4	2		0
10	341	146	68	33	33	16	8	8	8	4		1

the multiple eigenvalues for $(k, n) = (1, 7)$ and $(2, 14)$ respectively. We can see their rough structural similarity.

Let us see more detailed distributions for $k = 1, 2, \ldots, 5$ as given in Tables 11.1 to 11.5. The tables present the distribution of the non negative eigenvalues with multiplicities larger than or equal to 2. We have listed only non negative eigenvalues, where negative ones appear symmetrically for $k = 1$. For $k = 2, 3, 4, 5$ cases also, negative eigenvalues appear almost symmetrically on their multiplicities, except for a few values. Actually their monotonicity with respect to n holds. We also present the growth of the rates of $r^{(n;k)}$ for $k = 2$.

Observe the following structural similarity of $k = 2, 3, 4, 5$ cases to $k = 1$:

(1) The eigenvalues for $k \geq 2$, which are monotone increasing with respect to large n coincide with the ones for $k = 1$.

Table 11.2 Multiplicities of non-negative multiple eigenvalues for $k = 2$

n	$m_{1,2}^{(n;2)}$	$m_{1,3}^{(n;2)}$	$m_{1,4}^{(n;2)}$	$m_{1,5}^{(n;2)}$	$m_{2,5}^{(n;2)}$	$m_{1,6}^{(n;2)}$	$m_{1,7}^{(n;2)}$	$m_{2,7}^{(n;2)}$	$m_{3,7}^{(n;2)}$
1	0	0	0	0	0	0	0	0	0
2	1	0	0	0	0	0	0	0	0
3	0	0	0	0	0	0	0	0	0
4	4	0	0	0	0	0	0	0	0
5	6	0	0	0	0	0	0	0	0
6	22	3	0	0	0	0	0	0	0
7	42	6	0	0	0	0	0	0	0
8	104	21	3	0	0	0	0	0	0
9	210	50	6	0	0	0	0	0	0
10	460	118	24	3	3	0	0	0	0
11	930	252	54	6	6	0	0	0	0
12	1940	551	144	25	25	3	0	0	0
13	3906	1134	306	60	60	6	0	0	0
14	7966	2359	692	165	165	28	3	3	3
15	16002	4788	1434	366	366	66	6	6	6

Table 11.3 $k = 3$

n	$m^{(n;3)}\left(\frac{1}{4}\right)$	$m_{1,2}^{(n;3)}$	$m_{1,3}^{(n;3)}$	$m_{1,4}^{(n;3)}$	$m_{1,5}^{(n;3)}$	$m_{2,5}^{(n;3)}$
1	0	0	0	0	0	0
2	0	0	1	0	0	0
3	0	1	0	0	0	0
4	2	0	0	0	0	0
5	0	4	0	0	0	0
6	0	7	0	0	0	0
7	0	26	0	0	0	0
8	0	56	2	0	0	0
9	0	151	7	0	0	0
10	0	332	26	0	0	0
11	0	776	68	2	0	0
12	0	1653	196	7	0	0
13	0	3640	464	30	0	0
14	0	7604	1152	80	2	2
15	0	16157	2570	256	7	7

(2) One can find structural similarity of the distributions of the eigenvalues. Other multiple eigenvalues appear for every k steps for large n as is the case for $k = 1$. The order of appearance of the other multiple eigenvalues coincide. More concretely, for $k = 1$, other eigenvalues $\cos\frac{p}{n-1}\pi$ appear at the n-stage as multiple eigenvalues, and then they grow monotonically. For $k = 2$, they correspond to $\cos\frac{2p}{n}\pi$ with $n =$

Table 11.4	$k=4$			
n	$m^{(n;4)}\left(\frac{1}{5}\right)$	$m_{1,2}^{(n;4)}$	$m_{1,3}^{(n;4)}$	$m_{1,4}^{(n;4)}$
1	0	0	0	0
2	0	0	0	0
3	0	0	1	0
4	0	1	0	0
5	2	0	0	0
6	0	3	0	0
7	1	6	0	0
8	0	29	0	0
9	3	62	0	0
10	0	185	2	0
11	5	418	6	0
12	0	1061	31	0
13	9	2332	80	0
14	0	5427	265	2
15	15	11704	652	6

Table 11.5	$k=5$			
n	$m^{(n;5)}\left(\frac{1}{6}\right)$	$m^{(n;5)}\left(\frac{2}{6}\right)$	$m_{1,2}^{(n;5)}$	$m_{1,3}^{(n;5)}$
1	0	0	0	0
2	0	0	0	0
3	0	0	0	0
4	0	0	0	1
5	0	1	1	0
6	1	0	0	0
7	0	1	4	0
8	0	0	6	0
9	0	3	33	0
10	0	0	69	0
11	0	5	220	0
12	0	0	500	2
13	0	9	1333	6
14	2	0	3002	34
15	0	15	7327	93

$2, 4, 6, \ldots$. For general k, the eigenvalues are of the form $\cos \frac{p\pi}{\lfloor \frac{n-2}{k} \rfloor + 1}$, where $\lfloor \ \rfloor$ is the largest integer not greater than itself (Gauss symbol).

(3) In Tables 11.4 and 11.5, some of the eigenvalues are the extra ones which do not appear for the $k = 1$ case. They are included in the sets $\frac{\pm 1}{k+1}, \frac{\pm 2}{k+1}, \ldots$.

(4) The rates of the multiple eigenvalues $r^{(n;k)} = 2^{-n}\{ \sum_{i,j} m_{i,j}^{(n;k)} +$

$\sum_j m^{(n;k)}(\frac{\pm j}{k+1})$ } seem to grow to 1 with respect to n.

Based on these observations, we would like to propose the following:

Conjecture 11.1. *Let j be a non negative integer and let $\hat{Sp}(M^{(n)}_{k=j}) \subset$ $Sp(M^{(n)}_{k=j})$ be the set of multiple eigenvalues. Then:*

$$Sp(M^{(\lfloor (n-2)/j \rfloor)}_{k=1}) = \{1\} \cup \left(\hat{Sp}(M^{(n)}_{k=j}) \cap \hat{Sp}(M^{(n+1)}_{k=j}) \right) \quad for \quad n \geq 3.$$

Conjecture 11.2. *Let j be a non negative integer. There are $n_{\lambda,j} \in \mathbb{N}$ so that the following equalities hold:*

$$\lim_{n \to \infty} Sp\left(M^{(n)}_{k=1}\right) =$$
$$\lim_{n \to \infty} \{\lambda \in Sp(M^{(n)}_{k=j}) |\, 0 < m^{(n_{\lambda,j};j)}(\lambda) \leq \cdots \leq m^{(n-1;j)}(\lambda) \leq m^{(n;j)}(\lambda)\}.$$

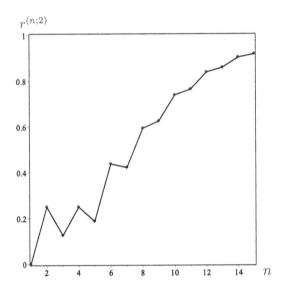

Fig. 11.3   Rates of the multiple eigenvalues $r^{(n;k)}$ for $k = 2$

So far we have found some similarity in the spectral distributions for various $k \geq 1$. It is quite unexpected for us to find any kind of structural similarity among BBS$_k$ and the lamplighter automaton, since BBS$_{k=1}$ is dynamically translation invariant, while BBS$_{k \geq 2}$ behaves essentially non linearly. It would be reasonable to expect to see more concrete dynamical similarity for several $k \geq 1$. On the other hand new eigenvalues are observed for $k \geq 2$, which may lead to essential differences of dynamics among BBS_k (see (3) above). Combining these opposite phenomena may lead to a much deeper understanding of BBS.

11.5   Stochastic matrices and ergodicity

Let us verify the stochastic property of the transition operators for BBS_k. We define a sequence of $k + 1$ matrices $(a_0^{(n)}, \ldots, a_k^{(n)})$ of 2^n sizes for $n = 0, 1, \ldots$ by the following matrix recursion (0 here represents a $2^n \times 2^n$ null matrix):

$$a_0^{(n+1)} = \begin{pmatrix} a_0^{(n)} & a_1^{(n)} \\ 0 & 0 \end{pmatrix}.$$

For $i = 1, \ldots, k - 1$

$$a_i^{(n+1)} = \begin{pmatrix} 0 & a_{i+1}^{(n)} \\ a_{i-1}^{(n)} & 0 \end{pmatrix}, \qquad a_k^{(n+1)} = \begin{pmatrix} 0 & 0 \\ a_{k-1}^{(n)} & a_k^{(n)} \end{pmatrix}$$

with the initial values $a_i^{(0)} = 1$ for all $i = 0, \ldots, k$.

We consider the approximations of the transition matrices given by $2^n \times 2^n$ matrices (see 11.1):

$$M_k^{(n)} = \frac{1}{2k + 2} \left(a_0^{(n)} + a_0^{(n)*} + \ldots + a_k^{(n)} + a_k^{(n)*} \right).$$

Lemma 11.2. $M_k^{(n)}$ *are doubly stochastic for all* $k \geq 1$, $n \geq 0$, *i.e. the sum of each row and each column is equal to 1.*

Proof. $M_k^{(n)}$ is symmetric and therefore it suffices to verify that the sum of the columns is constant.

It follows from the recursive relations for $a_0^{(n+1)}, \ldots, a_k^{(n+1)}$ that the sum of the columns of each $a_i^{(n+1)}$ is constant and equal to 1 for $1 \leq i \leq k$.

Thus it is enough to show that $a_0^{(n)*} + \ldots + a_k^{(n)*}$ has a constant column sum. Let us prove this by induction. It is clear for $n = 0$. Then using recursion formula

$$a_0^{(n+1)*} + \cdots + a_k^{(n+1)*} = \begin{pmatrix} a_0^{(n)*} & a_{k-1}^{(n)*} + \cdots + a_0^{(n)*} \\ a_1^{(n)*} + \cdots + a_k^{(n)*} & a_k^{(n)*} \end{pmatrix}$$

and thus the sum of the left matrix blocks and right matrix blocks is equal to

$$a_0^{(n)*} + \ldots + a_k^{(n)*}.$$

Therefore the statement follows by induction. □

Recall that a Mealy automaton is twisted invertible, if the pair of the transition and exist functions $(\phi, \psi) : Q \times S \cong Q \times S$ are pairwisely invertible (see 6.2.3 and 7.4). The above lemma also follows from the following:

Proposition 11.2. *Suppose a Mealy automaton is twisted invertible. Then the approximation of the Markov operators consists of all doubly stochastic matrices.*

Proof. Let $M^{(n)} = \frac{1}{2k+2}(a_0^{(n)} + a_0^{(n)*} + \ldots + a_k^{(n)} + a_k^{(n)*})$ be the approximation of the Markov operator for $n \geq 1$. Actually it follows from invertibility that the sum of each column for $A^{(n)} = \frac{1}{k+1}(a_0^{(n)} + \ldots + a_k^{(n)})$ is constant and equal to 1. Since the pair of the functions (ϕ, ψ) gives permutation on the product of the set of the states with each level set of the tree, the adjoint $(A^{(n)})^* = \frac{1}{k+1}((a_0^{(n)})^* + \ldots + (a_k^{(n)})^*)$ also satisfies the same property.

Then the conclusion follows since $M^{(n)}$ is twice the sum of $A^{(n)}$ with $(A^{(n)})^*$. □

11.6 Ergodicity on the boundary of the binary tree

Let $\{M_1^{(n)}\}_{n=1,2,\ldots}$ be the approximation of the transition operators for lamplighter or BBS$_{k=1}$ automata. We have verified that those are stochastic $2^n \times 2^n$ matrices equipped with the canonical maps:

$$\cdots \to M_1^{(n+1)} \to M_1^{(n)} \to \cdots$$

Definition 11.2. Let M be a stochastic $l \times l$ matrix. M is ergodic, if there is $s_0 \geq 1$ and $\alpha > 0$ so that inequalities:

$$m_{i,j}^{(s_0)} \geq \alpha$$

hold for all i, j, where $m_{i,j}^s$ are the components of $M^s = (m_{i,j}^{(s)})_{1 \leq i,j \leq l}$.

For a stochastic matrix, if the above property is satisfied for some s_0, then the same property holds for all $s \geq s_0$.

Recall the fundamental result on ergodicity (see [S]):

Theorem 11.3. *Let M be a stochastic $l \times l$ matrix, and consider the associated Markov chain on the space $X = \{1, \ldots, l\}$. If M is ergodic, then there is a unique probability distribution π on X which satisfies two properties* (1) $\pi M = \pi$, *and* (2) $\lim_{s \to \infty} m_{i,j}^{(s)} = \pi_j$.

The unique probability distribution $\pi = (\pi_1, \ldots, \pi_k)$ is called the *stationary distribution* with respect to M.

Corollary 11.1. *Let $M_L^{(n)}$ and $M_B^{(n)}$ be the approximation of the transition operators for the lamplighter and $BBS_{k=1}$ automata respectively.*
Then they are all ergodic.

Proof. The result follows from our computation of their spectra in theorem 11.2 with Lemma 11.3 below. □

Lemma 11.3. *M is ergodic, if and only if the spectrum of M satisfies the following:*
(1) the multiplicity of the eigenvalue 1 is just 1, and
(2) it does not contain -1.

Proof. Suppose M is ergodic. Let v_1 and v_2 be two orthogonal eigenvectors with eigenvalue 1. Then $v_i M = v_i$ holds, and so:

$$\langle v_1 M^{2s}, v_2 \rangle = \langle v_1 M^s, v_2 M^s \rangle = \langle v_1, v_2 \rangle = 0$$

must hold. Let a_i be the sum of coordinates of v_i. Then a_i cannot be zero, since $v_i = \lim_{s \to \infty} v_i M^s = a_i \pi$ holds by theorem 11.3. By letting $s \to \infty$ in the above equalities, it follows that $\pi = 0$ is a zero vector, which is a contradiction, since $\pi_i \geq \alpha > 0$. So the multiplicity of the eigenvalue 1 must be less than or equal to 1. It is at least 1 because constant vectors have eigenvalue 1.
Similarly the limit exists:

$$wM^s \equiv (x_1, \ldots, x_n)M^s \to (a\pi_1, \ldots, a\pi_n)$$

by theorem 11.3, where $a = \sum_{i=1}^{n} x_i$. But if w is an eigenvector with eigenvalue -1, then $wM^s \in \{w, -w\}$ oscillates, which is a contradiction.
Conversely suppose the above two properties hold. Let $\{v_1, \ldots, v_k\}$ be the orthogonal eigenvectors such that $v_1 = (1, \ldots, 1)$ corresponds to the eigenvalue 1. Then for any $v = \sum_{i=1}^{k} a_i v_i$,

$$\lim_{s \to \infty} vM^s = a_1 v_1 + \lim_{s \to \infty} \sum_{i=2}^{k} \lambda_i^s a_i v_i = a_1 v_1$$

holds, since $-1 < \lambda_i < 1$ for $i \geq 2$.

If M were not ergodic so that there exist i, j where $m_{i,j}^{(s)} = 0$ holds for every s, then $\langle \delta_i M^s, \delta_j \rangle = m_{i,j}^{(s)} = 0$ holds for $\delta_i = (0, \dots, 0, 1, 0, \dots)$. Then there exist i, j such that $\langle \delta_i M^l, \delta_j \rangle = 0$ holds for infinitely many l. So it also holds for $l \to \infty$. It follows that δ_i or δ_j is orthogonal to $v_1 = (1, \dots, 1)$, which is a contradiction. $\square$

Remark 11.1. For the stochastic matrix, property (1) is equivalent to connectivity, and property (2) is equivalent to non bi-partiteness of the associated graph.

Countable state ergodic Markov chains is an important subject in relation with statistic mechanics. It follows from corollary 11.1 below that approximation of BBS transition operator $M_{k=1}^{(n)}$ gives the ergodic Markov chains over the set of paths:
$$\Omega(n) = \{(w_1, w_2, \dots) \mid w_i \in \{1, \dots, 2^n\}\}$$
with the unique ergodic distributions $\pi^{(n)} = (\pi_1^{(n)}, \dots, \pi_{2^n}^{(n)})$.

Let μ^n be the probability measures on $\Omega(n)$ with respect to the transition operators $M^{(n)}$ of BBS_k for any $k \geq 1$ or the lamplighter and Burnside automata. One may expect that these families of Markov chains defined by $\{M^{(n)}\}_{n=1}^{\infty}$ can give countable state Markov chains over the path space:
$$\Omega(\infty) = \{(w_1, w_2, \dots) \mid w_i \in \mathbb{N}\}.$$
At present their structures have not yet been analyzed.

From the viewpoint of scale transform, the countable Markov chains lie in the ultimate scale, since the process is on the boundary of the tree which is located at the infinity of automata dynamics, while such automata come from PDE passing through rational dynamics.

References

Computation of the spectra of the Markov operator on the lamplighter group was done in [GZ]. Spectral coincidence between the Markov operators on the lamplighter group and the BBS semi-group was found in [KTZ]. Spectral analysis on automata groups was given in terms of fractal groups in [BG]. A criterion of amenability on automata groups was given in [BKN]. A basic theory on random walk on infinite groups is given in [Woe]. A stochastic matrix gives the Markov chains on path spaces. See [Sin] and also [Kit] on symbolic dynamics.

Chapter 12

Rough comparison between various differential equations

Tropical geometry is a kind of scale transform between dynamical systems, which provides a correspondence between automata and real rational dynamics. It allows us to treat rational dynamics as a class such that equivalent rational dynamics share the same analytic behaviors on the large scale in some sense.

There has been the extended development of partial differential equations using approximations by real rational dynamics. Combining such approximations with tropical geometry will lead us to study a comparison of solutions to different PDE systems. Namely from the viewpoint of dynamical scaling limits, automata can be regarded as a frame of dynamical systems, which plays the role of underlying mechanisms for various systems of partial differential equations.

Let us consider which kinds of framework of PDE systems we can pick up. For example a combination of these two aspects can create a new connection between geometric group theory and partial differential equations. If a PDE system is reduced to an invertible Mealy automaton, then one can say that its framework, or underlying dynamics is invertible (!) even if solutions themselves to the PDE system cannot be reversed in time. There are characteristic phenomena for each theory in automata, groups and dynamical systems, for example calculating machine, freeness, recursivity, ergodicity, amenability, etc. In particular the notions which appear in geometric group theory are quite important for us, since the properties are invariant with respect to quasi-isometry. It would be of particular interest for us to translate these notions into PDE systems and compare them; this will be left as a future research direction.

The key analytic tool is *dynamical inequalities* which we develop in a general form. This is essentially required when one uses approximation of

PDE systems by rational dynamics, since one has to estimate *error terms* which appear during such a process.

Mealy automata correspond to the hyperbolic systems of first order PDE of 2-variables. We develop a basic analysis of the class of PDE systems, which includes existence, uniqueness, C^1 estimates with explicit estimates on the constants, and so on. It allows us to produce solutions to the PDE systems which can be applied to the asymptotic comparison theorem which we will describe below.

12.1   Formal Taylor expansion and ODE

We introduce a procedure to approximate discrete dynamics defined by elementary rational functions of n variables by a partial differential equation.

Let $u : (0, \infty) \to (0, \infty)$ be a $C^{\alpha+1}$ function. For $1 \le |i| \le n - 1$, let us take the Taylor expansions around $x \in (0, \infty)$:

$$u(x + i\epsilon) = u(x) + i\epsilon u_x + \frac{(i\epsilon)^2}{2} u_{2x} + \cdots + \frac{(i\epsilon)^\alpha}{\alpha!} u_{\alpha x} + \frac{(i\epsilon)^{(\alpha+1)}}{(\alpha + 1)!} u_{(\alpha+1)x}(\xi_i)$$

for small $|\epsilon| << 1$, where:

$$\begin{cases} x \le \xi_i \le x + i\epsilon, & i \ge 0 \\ x + i\epsilon \le \xi_i \le x & i < 0. \end{cases}$$

Let $f = \frac{k}{h} : \mathbb{R}^n_{>0} \to (0, \infty)$ be an elementary rational function, where h and k are both elementary polynomials. Let us consider the discrete dynamics defined by the equation:

$$z_{N+1} = f(z_{N-n+1}, \ldots, z_N).$$

For $N = 0, 1, 2, \ldots$, let us introduce the change of variables:

$$z_N \equiv \epsilon u(\epsilon N) = \epsilon u(x), \quad \left(N = \frac{x}{\epsilon} \right)$$

and consider the difference:

$$z_{N+1} - f(z_{N-n+1}, \ldots, z_N) = \epsilon u(x + \epsilon) - f(\epsilon u(x - (n-1)\epsilon), \ldots, \epsilon u(x)).$$

Let us insert the Taylor expansions:

$$= \epsilon(u + \epsilon u_x + \frac{\epsilon^2}{2} u_{2x} + \ldots)$$
$$- f(\epsilon(u - (n-1)\epsilon u_x + \frac{(n-1)^2 \epsilon^2}{2} u_{2x} + \ldots), \ldots, \epsilon u).$$

By reordering the expansions with respect to the exponents of ϵ, there are rational numbers $a_0, a_1, \cdots \in \mathbb{Q}$ so that the equality holds:

$$\epsilon u(x + \epsilon) - f(\epsilon u(x - (n-1)\epsilon), \dots, \epsilon u(x))$$

$$= \frac{\epsilon a_0 u + \epsilon^2 a_1 u_x + \epsilon^3 a_2 u u_x + .. + \epsilon^{\alpha+1} a_s u_{\alpha x} + \epsilon^{\alpha+2} a_{s+1} u_{(\alpha+1)x}(\xi) + ..}{h(\epsilon u(x - (n-1)\epsilon), \dots, \epsilon u(x))}$$

$$\equiv \frac{\epsilon F^1(u) + \epsilon^2 F^2(u_x) + \epsilon^3 F^3(u, u_x) + .. + \epsilon^m F^{m'}(u, .., u_{(\alpha+1)x}(\xi)) + ..}{h(\epsilon u(x - (n-1)\epsilon), \dots, \epsilon u(x))}$$

where F^k are monomials.

Let us choose finite subsets $A \subset \{1, 2, 3, \dots\}$, and divide the expanded sum into two terms as:

$$= \frac{\Sigma_{i \in A} \ \epsilon^i F^{i'}(u, u_x, \dots, u_{\alpha x})}{h(\epsilon u(x - (n-1)\epsilon), \dots)}$$

$$+ \frac{\Sigma_{j \in A^c} \ \epsilon^j F^{j'}(u, u_x, \dots, u_{(\alpha+1)x}(\xi))}{h(\epsilon u(x - (n-1)\epsilon), ..)}$$

$$\equiv \mathbf{F}(\epsilon, u, u_x, \dots, u_{\alpha x})$$

$$+ \epsilon^2 \mathbf{F}^1(\epsilon, u, u_x, .., u_{(\alpha+1)x}(\xi_1), .., u_{(\alpha+1)x}(\xi_{n-1})).$$

We always choose A so that two conditions are satisfied;

(1) $\mathbf{F}$ does not contain the term $u_{(l+1)x}(\xi)$,
(2) $1 \in A$, i.e. F^1 is included in $\mathbf{F}$.

Remark 12.1. In many concrete cases later, we choose elementary rational functions and A so that the corresponding F^1 vanishes.

Now fix $\epsilon > 0$, and suppose u satisfies the equation:

$$\mathbf{F}(\epsilon, u, u_x, \dots, u_{\alpha x}) = 0.$$

Then the difference satisfies the equality:

$$\epsilon u(x + \epsilon) - f(\epsilon u(x - (n-1)\epsilon), \dots, \epsilon u(x)) = \epsilon^2 \mathbf{F}^1(\epsilon, u, u_x, \dots).$$

We say that $\mathbf{F}$ is the *leading term*, and $\mathbf{F}^1$ the error term for u respectively.

Remark 12.2. Conversely when one starts from the ordinary differential equation $\mathbf{F}(\epsilon, u, u_x, u_{\alpha x}) = 0$, one will have quite flexible choices of relative elementary functions f and A with the leading term $\mathbf{F}$. Various choices of f will produce different error terms $\mathbf{F}^1$. So a 'better' choice of f will give us 'better' estimates of solutions $\mathbf{F}(\epsilon, u, u_x, \dots, u_{\alpha x}) = 0$ in terms of rational dynamics by f.

Let us define ϵ *variation* of $\mathbf{F}^1$ by

$$||\mathbf{F}^1(\epsilon, u, u_x, \ldots, u_{\alpha x}, u_{(\alpha+1)x}(\xi_1), \ldots, u_{(\alpha+1)x}(\xi_{n-1}))||_\epsilon(x) \equiv$$

$$\sup_{\mu_i - x \in [-n\epsilon, 0]} |\mathbf{F}^1(\epsilon, u(x-\epsilon), .., u_{\alpha x}(x-\epsilon), u_{(\alpha+1)x}(\mu_1), .., u_{(\alpha+1)x}(\mu_{n-1}))|$$

where n is the number of the variables of f.

Let us say that a $C^{\alpha+1}$ function $u : (0, \infty) \to (0, \infty)$ is ϵ_0 *controlled*, if there is some constant $C > 0$ so that the ϵ_0 variation of $\mathbf{F}^1$ satisfy the pointwise estimates for all $x \in (0, \infty)$:

$$Cu(x) \geq ||\mathbf{F}^1(\epsilon, u, u_x, \ldots, u_{\alpha x}, u_{(\alpha+1)x}(\xi_1), \ldots, u_{(\alpha+1)x}(\xi_{n-1}))||_{\epsilon_0}(x).$$

12.2 Comparison of solutions for ODE

Let $g = \frac{d}{e}$ be another elementary rational function which is tropically equivalent to f.

Let $v : (0, \infty) \to (0, \infty)$ be another $C^{\alpha+1}$ function. By replacing f by g and choosing other subsets $B \subset \{1, 2, 3, \ldots\}$, one has its leading and error terms $\mathbf{G}$ and $\mathbf{G}^1$ respectively:

$$\epsilon v(x + \epsilon) - g(\epsilon v(x - (n-1)\epsilon), \ldots, \epsilon v(x))$$

$$= \frac{\Sigma_{i \in B} \ \epsilon^i G^{i'}(v, v_x, \ldots, v_{\alpha x})}{e(\epsilon v(x - (n-1)\epsilon), \ldots)} + \frac{\Sigma_{j \in B^c} \ \epsilon^j G^{j'}(v, v_x, \ldots, v_{(\alpha+1)x}(\xi'))}{e(\epsilon v(x - (n-1)\epsilon), \ldots)}$$

$$\equiv \mathbf{G}(\epsilon, v, \ldots, v_{\alpha x}) + \epsilon^2 \mathbf{G}^1(\epsilon, v, \ldots, v_{\alpha x}, v_{(\alpha+1)x}(\xi_1'), \ldots, v_{(\alpha+1)x}(\xi_{n-1}')).$$

Let us fix a small $\epsilon > 0$, and suppose one has two positive solutions $u, v : (0, \infty) \to (0, \infty)$ to the equations:

$$\mathbf{F}(\epsilon, u, u_x, \ldots) = 0,$$
$$\mathbf{G}(\epsilon, v, v_x, \ldots) = 0.$$

What we want to do is to estimate the ratios:

$$\left(\frac{u(x)}{v(x)}\right)^{\pm 1} = \left\{\frac{u(x)}{v(x)}, \frac{v(x)}{u(x)}\right\}.$$

Notice that u and v belong to different ODEs and hence one will not be able to analyze their behaviors directly from the equations.

For this we introduce the *initial rates*:

$$[u : v]_\epsilon \equiv \sup_{x \in (0, \epsilon]} \left(\frac{u(x)}{v(x)}\right)^{\pm 1}.$$

Recall that associated with f are the Lipschitz constant $c_f \geq 1$ and the number of the components M_f. Let $c = \max(c_f, c_g)$ and $M = \max(M_f, M_g)$.

Theorem 12.1. *Let f and g be both elementary rational functions of n variables, which are mutually tropically equivalent. Let $\mathbf{F}$ and $\mathbf{G}$ be their leading terms of order at most $\alpha \geq 0$.*

Then there are constants a, b so that for any positive $C^{\alpha+1}$ solution $u, v : (0, \infty) \to (0, \infty)$ to the equations:

$$\mathbf{F}(\epsilon, u, u_x, \ldots, u_{\alpha x}) = 0,$$
$$\mathbf{G}(\epsilon, v, v_x, \ldots, v_{\alpha x}) = 0$$

the following holds.

Assume both u and v are ϵ_0 controlled bounded by C. Then for any $0 < \epsilon \leq \min(\frac{1}{2C}, \epsilon_0)$, the estimates hold:

$$\left(\frac{u(x)}{v(x)}\right)^{\pm 1} \leq \exp(a^{\epsilon^{-1}x+1} + b^{\epsilon^{-1}x+1} \log[u : v]_{n\epsilon}).$$

The proof is given in sec. 12.3 below.

Remark 12.3. (1) Notice that both a, b are independent of C, but $\epsilon > 0$ depends on C^{-1}.

(2) Actually we can specify a, b explicitly. Later we will describe them.

Example 12.1. Let us consider a simple equation:

$$\mathbf{F}(u, u_x) = u_x + u^2 = 0.$$

It has solutions $u(x) = \frac{a}{1+ax}$ with the initial values $u(0) = a > 0$. Let $z_N = \epsilon u(x)$ with $x = N\epsilon$, and take the Taylor expansion $\epsilon u(x + \epsilon) = \epsilon u(x) + \epsilon^2 u_x(x) + \frac{\epsilon^3}{2} u_{2x}(\xi)$. We choose the elementary rational function $f(x) = x(1 + x)^{-1}$ and calculate the difference:

$$u(x + \epsilon) - f(\epsilon u(x))$$
$$= \epsilon^2 \frac{u_x + u^2}{1 + \epsilon u}(x) + \epsilon^3 \frac{\frac{1}{2}u_{2x}(\xi) + u(x)u_x(x) + \frac{\epsilon}{2}u(x)u_{2x}(\xi)}{1 + \epsilon u(x)}.$$

Thus u is ϵ_0 controlled, since $1 + \epsilon u(x) \geq 1$, and the estimates $|u_{2x}(\xi)|, |uu_x(x)|, |u(x)u_{2x}(\xi)| \leq Cu(x)$ hold uniformly in $x \in (0, \infty)$ for some $C = C(a, \epsilon_0) \geq 0$ and $|x - \xi| \leq \epsilon$.

The corresponding $(\max, +)$-function to f is given by $x_{N+1} = x_N - \max(0, x_N)$. Notice the equality:

$$x_N - \max(0, x_N) = x_N - \max(0, x_N, x_N).$$

The tropical inverse for the latter is given by $g(y) = y(1+2y)^{-1}$. By choosing the same scaling parameter, one obtains the leading term $\mathbf{G}(v, v_x) = v_x + 2v^2$, which has solutions $v(x) = \frac{a'}{2a'x+1}$. Their ratio is in fact uniformly bounded:

$$\left(\frac{u(x)}{v(x)}\right)^{\pm} = \left(\frac{a(2a'x+1)}{a'(ax+1)}\right)^{\pm 1} \leq 2\left(\frac{a}{a'}\right)^{\pm 1} \leq 2[u:v]_\epsilon.$$

12.3 Dynamical inequalities

The key analytic tool to induce the estimates on solutions in theorem 12.1 is comparison of *dynamical inequalities* in tropical geometry. Let us describe it in detail below.

Let φ^1 and φ^2 be relative $(\max, +)$-functions with the number of the components $M = \max(M_{\varphi^1}, M_{\varphi^2})$.

Suppose they are mutually equivalent and so the equality $c = c_{\varphi^1} = c_{\varphi^2}$ of the Lipschitz constants holds. Let us denote tropically equivalent functions by f_t^1 and f_t^2 correspondingly.

Let us analyze orbits which admit dynamical inequalities. Take an initial sequence $(z_0, z_1, \ldots, z_{n-1}) \in \mathbb{R}^n_{>0}$, and consider the orbits $\{z_i^l\}_i$ to the same initial values $z_i^l = z_i$ for $0 \leq i \leq n - 1$:

$$z_N^l = f_t^l(z_{N-n+1}^l, \ldots, z_{N-1}^l).$$

Example 12.2. Suppose that another sequence $\{w_i\}_i$ satisfies the dynamical inequality:

$$f_t^1(w_{N-n+1}, \ldots, w_{N-1}) \leq w_N \leq f_t^2(w_{N-n+1}, \ldots, w_{N-1})$$

for all $N \geq n$, with the same initial value $w_i = z_i$ for $0 \leq i \leq n - 1$.

Then the uniform estimates hold:

$$\left(\frac{w_i}{z_i^1}\right)^{\pm 1}, \ \left(\frac{w_i}{z_i^2}\right)^{\pm 1} \leq M^{2P_i(c)}$$

where $P_i(c) = \frac{c^{i-n+1}-1}{c-1}$.

Remark 12.4. The above assumption implies that $f_t^1 \leq f_t^2$ should hold at least on the orbits. Notice that there are some cases where the inequalities $f_t^1 \leq f_t^2$ hold. For example if φ^2 has the presentation as $\varphi^2 = \max(\varphi^1, \varphi^1)$, then $f_t^1 \leq f_t^2 = 2f_t^1$ holds.

Proof. Let $x_i = \log_t w_i$ and $y_i^l = \log_t z_i^l$. Then the equations hold by lemma 3.2:

$$y_N^l = \varphi_t^l(y_{N-n+1}^l, \ldots, y_{N-1}^l).$$

Since $\log_t$ is increasing, the estimates hold:

$$\varphi_t^1(x_{N-n+1}, \ldots, x_{N-1}) \leq x_N \leq \varphi_t^2(x_{N-n+1}, \ldots, x_{N-1}).$$

Now we claim that the estimates:

$$|x_i - y_i^1| \leq 2P_i(c) \log_t M$$

hold for all $i \geq 0$. Let $\bar{x}_i = (x_i, \ldots, x_{i+n-1})$.

For $i = 0$, $x_0 = y_0^1$ holds, since they have the same initial values.

Take any $i \geq 0$, and divide into two cases. Firstly suppose the inequality $x_{i+1} \geq y_{i+1}^1$ holds. Then the estimates hold by lemma 3.3 and 3.4:

$$\begin{aligned}
0 \leq x_{i+1} - y_{i+1}^1 &\leq \varphi_t^2(\bar{x}_{i-n}) - \varphi_t^1(\bar{y}_{i-n}) = |\varphi_t^2(\bar{x}_{i-n}) - \varphi_t^1(\bar{y}_{i-n})| \\
&\leq |\varphi_t^2(\bar{x}_{i-n}) - \varphi^2(\bar{x}_{i-n})| \\
&\quad + |\varphi^2(\bar{x}_{i-n}) - \varphi^1(\bar{y}_{i-n})| + |\varphi^1(\bar{y}_{i-n}) - \varphi_t^1(\bar{y}_{i-n})| \\
&\leq 2\log_t M + c \max_{0 \leq m \leq i} |x_m - y_m^1|.
\end{aligned}$$

Conversely suppose $x_{i+1} \leq y_{i+1}^1$ holds. Then the estimates hold:

$$\begin{aligned}
0 \leq y_{i+1}^1 - x_{i+1} &\leq \varphi_t^1(\bar{y}_{i-n}) - \varphi_t^1(\bar{x}_{i-n}) = |\varphi_t^1(\bar{y}_{i-n}) - \varphi_t^1(\bar{x}_{i-n})| \\
&\leq |\varphi_t^1(\bar{y}_{i-n}) - \varphi^1(\bar{y}_{i-n})| \\
&\quad + |\varphi^1(\bar{y}_{i-n}) - \varphi^1(\bar{x}_{i-n})| + |\varphi^1(\bar{x}_{i-n}) - \varphi_t^1(\bar{x}_{i-n})| \\
&\leq 2\log_t M + c \max_{0 \leq m \leq i} |x_m - y_m^1|.
\end{aligned}$$

Thus the estimates $|x_{i+1} - y_{i+1}^1| \leq 2\log_t M + c \max_{0 \leq m \leq i} |x_m - y_m^1|$ hold in any case. By iteration,

$$\begin{aligned}
|x_{i+1} - y_{i+1}^1| &\leq 2\log_t M + c \max_{0 \leq m \leq i} |x_i - y_i^1| \\
&\leq 2\log_t M + c\{2\log_t M + c \max_{0 \leq m \leq i-1} |x_m - y_m^1|\} \\
&= 2(1 + c)\log_t M + c^2 \max_{0 \leq m \leq i-1} |x_m - y_m^1| \\
&\leq \cdots \leq 2P_{i+1}(c) \log_t M + c^{i-n+2} \max_{0 \leq m \leq n-1} |x_m - y_m^1| = 2P_{i+1}(c) \log_t M
\end{aligned}$$

holds. This verifies the claim.

The left hand side is equal to $\log_t(\frac{w_{i+1}}{z_{i+1}^1})^{\pm 1}$ and the right hand side is equal to $\log_t M^{2P_{i+1}(c)}$. Since $\log_t$ is distance increasing, we obtain the estimates:

$$\left(\frac{w_i}{z_i^1}\right)^{\pm 1} \leq M^{2P_i(c)}.$$

The estimates $(\frac{w_i}{z_i'})^{\pm 1} \leq M^{2P_i(c)}$ are obtained the same way, and we omit repetition. □

Now we induce the main estimates:

Proposition 12.1. *Let us take four elementary rational functions f_t, f_t', g_t, g_t' which are all tropically equivalent.*
Let $\{v_N\}_N$ and $\{u_N\}_N$ be positive sequences which satisfy the estimates:

$$g_t'(v_{N-n}, \ldots, v_{N-1}) \leq v_N \leq g_t(v_{N-n}, \ldots, v_{N-1}),$$
$$f_t'(u_{N-n}, \ldots, u_{N-1}) \leq u_N \leq f_t(u_{N-n}, \ldots, u_{N-1})$$

for all $N \geq n$. Then the ratios satisfy the uniform estimates:

$$\left(\frac{v_N}{u_N}\right)^{\pm 1} \leq M^{8P_N(c)} \Big[\max_{0 \leq i \leq n-1} \left(\frac{u_i}{v_i}\right)^{\pm 1} \Big]^{c^{N-n}}$$

where $c = c_f = c_{f'} = c_g = c_{g'}$ and $M = \max(M_f, M_{f'}, M_g, M_{g'})$.

Proof. Let us consider two orbits $\{z_N\}_N$ and $\{z_N'\}_N$ which are both defined by the same equation $z_N = g_t(z_{N-n}, \ldots, z_{N-1})$ and $z_N' = g_t(z_{N-n}', \ldots, z_{N-1}')$ but with the different initial values $z_i = v_i$ and $z_i' = u_i$ for $0 \leq i \leq n - 1$ respectively. □

Assertion 12.1. The estimates hold:

$$\left(\frac{z_N}{z_N'}\right)^{\pm 1} \leq M^{2P_i(c)}[z_i; z_i']^{c^{i-n}}$$

where $[z_i; z_i'] = \max_{0 \leq i \leq n-1}(\frac{z_i}{z_i'})^{\pm 1}$.

Proof. (see lemma 6.5)
   Step 1: Let φ and φ_t be the tropical correspondences to g_t, and let $\{x_i\}_i$ and $\{y_i\}_i$ be other orbits of φ with their initial values $x_i = \log_t v_i$ and $y_i = \log_t u_i$ for $0 \leq i \leq n - 1$ respectively. We claim that the estimates:

$$|x_i - y_i| = |\varphi(\bar{x}_{i-n}) - \varphi(\bar{y}_{i-n})| \leq c^{i-n} \log_t[z_i; z_i']$$

hold. In fact we have the estimates:

$$|x_{i+1} - y_{i+1}| = |\varphi(\bar{x}_{i-n}) - \varphi(\bar{y}_{i-n})| \leq c \max_{i-n \leq m \leq i-1} |x_m - y_m|$$
$$\leq c^2 \max_{i-n-1 \leq m \leq i-2} |x_m - y_m| \leq \ldots$$
$$\leq c^{i-n+1} \max_{0 \leq m \leq n-1} |x_m - y_m| = c^{i-n+1} \log_t[z_i; z_i'].$$

Step 2: It follows by lemma 3.5 that the estimates:

$$|x_i - x_i'|, |y_i - y_i'| \leq P_i(c) \log_t M$$

hold where $x_i' = \log_t z_i$ and $y_i' = \log_t z_i'$ respectively.

With step 1, we obtain the estimates:

$$\log_t(\frac{z_N}{z_N'})^{\pm 1} = |x_i' - y_i'| \leq |x_i - x_i'| + |x_i - y_i| + |y_i - y_i'|$$

$$\leq 2P_i(c) \log_t M + c^{i-n} \log_t [z_i; z_i']$$

$$= \log_t \{M^{2P_i(c)}[z_i; z_i']^{c^{i-n}}\}.$$

Since $\log_t$ is monotone increasing, we obtain the desired estimates. □

Proof of proposition 12.1. Let $p_i = f_t(p_{i-n}, \ldots, p_{i-1})$ be another orbit with the initial value $p_i = u_i$ for $0 \leq i \leq n - 1$.

It follows from theorem 3.1 that the estimate holds.

$$(\frac{p_i}{z_i'})^{\pm 1} \leq M^{2P_i(c)}.$$

Now we have the estimates:

$$(\frac{u_i}{v_i})^{\pm 1} = (\frac{u_i}{p_i})^{\pm 1}(\frac{p_i}{z_i'})^{\pm 1}(\frac{z_i'}{z_i})^{\pm 1}(\frac{z_i}{v_i})^{\pm 1}$$

$$\leq M^{2P_i(c)} M^{2P_i(c)} M^{2P_i(c)} [z_i; z_i']^{c^{i-n}} M^{2P_i(c)}$$

$$= M^{8P_i(c)}[z_i; z_i']^{c^{i-n}}.$$

This completes the proof of proposition 12.1.

Proof of theorem 12.1. We verify the estimates:

$$(\frac{u(x)}{v(x)})^{\pm 1} \leq (2M)^{8\frac{c^{\epsilon^{-1}x-n+1}-1}{c-1}} ([u:v]_{n\epsilon})^{c^{\epsilon^{-1}x-n}}.$$

Let f and g be both n variables, and $(\mathbf{F}, \mathbf{F}^1)$ and $(\mathbf{G}, \mathbf{G}^1)$ be pairs of leading and error terms respectively.

Let us choose $0 < \epsilon \leq \min(\frac{1}{2C}, \epsilon_0)$. By the assumption, the pointwise estimates hold:

$$Cu(x+\epsilon) \geq |\mathbf{F}^1(\epsilon, u(x), u_x(x), \ldots, u_{\alpha x}(x), u_{(\alpha+1)x}(\xi_1), \ldots, u_{(\alpha+1)x}(\xi_{n-1}))|.$$

In particular the estimates $\epsilon^2 |\mathbf{F}^1| \leq \frac{1}{2C}\epsilon|\mathbf{F}^1| \leq \frac{1}{2}\epsilon u(x + \epsilon)$ hold.

Let us consider the equalities:

$$\epsilon u(x + \epsilon) - f(\epsilon u(x - (n-1)\epsilon), \ldots, \epsilon u(x))$$

$$= \mathbf{F}(\epsilon, u, \ldots, u_{\alpha x}) + \epsilon^2 \mathbf{F}^1(\epsilon, u, \ldots, u_{\alpha x}, u_{(\alpha+1)x}(\xi_1), \ldots, u_{(\alpha+1)x}(\xi_{n-1}))$$

$$= \epsilon^2 \mathbf{F}^1(\epsilon, u, u_x, \ldots, u_{\alpha x}, u_{(\alpha+1)x}(\xi_1), \ldots, u_{(\alpha+1)x}(\xi_{n-1}))$$

since u obeys the equation $\mathbf{F}(\epsilon, u, u_x, \dots) = 0$.

Then combining with the above inequality, one obtains the estimates:

$$\frac{1}{2}f(\epsilon u(x - (n-1)\epsilon), \dots, \epsilon u(x)) \le \epsilon u(x + \epsilon)$$
$$\le 2f(\epsilon u(x - (n-1)\epsilon), \dots, \epsilon u(x)).$$

By the same way one obtains the estimates by replacing f by g:

$$\frac{1}{2}g(\epsilon v(x - (n-1)\epsilon), \dots, \epsilon v(x)) \le \epsilon v(x + \epsilon)$$
$$\le 2g(\epsilon v(x - (n-1)\epsilon), \dots, \epsilon v(x)).$$

Both $\frac{1}{2}f$ and $2f$ are tropically equivalent to f, and hence $\frac{1}{2}f, 2f, \frac{1}{2}g, 2g$ are all tropically equivalent. Notice that the number of the components of $\frac{1}{2}f$ and $2f$ are both $2M_f$.

Let $\epsilon u(N\epsilon) = u_N$ and $\epsilon v(N\epsilon) = v_N$. Then we have the estimates:

$$\frac{1}{2}f(v_{N-n}, \dots, v_N) \le v_{N+1} \le 2f(v_{N-n}, \dots, v_N),$$
$$\frac{1}{2}g(u_{N-n}, \dots, u_N) \le u_{N+1} \le 2g(u_{N-n}, \dots, u_N).$$

Thus the estimates hold by proposition 12.1:

$$\left(\frac{u(N\epsilon)}{v(N\epsilon)}\right)^{\pm 1} \le (2M)^{8P_N(c)} \sup_{0 \le i \le n-1} \left(\frac{u(\epsilon i)}{v(\epsilon i)}\right)^{\pm c^{N-n}}$$
$$\le (2M)^{8P_N(c)}([u:v]_{(n-1)\epsilon})^{c^{N-n}}.$$

For any $0 \le \mu \le 1$, let us consider their translations $u(x + \mu)$ and $v(x + \mu)$. If we replace all the above estimates for $u(x\epsilon)$ by $u((x+\mu)\epsilon)$ and $v(x\epsilon)$ by $v((x+\mu)\epsilon)$, then one obtains the estimates:

$$\left(\frac{u(N\epsilon + \mu)}{v(N\epsilon + \mu)}\right)^{\pm 1} \le (2M)^{8P_N(c)}[u:v]_{n\epsilon}^{c^{N-n}}.$$

$\cup_{N \ge 1}(N + \mu)\epsilon$ covers all the points $x \in (0, \infty)$, and so the estimates hold:

$$\left(\frac{u(x)}{v(x)}\right)^{\pm 1} \le (2M)^{8\frac{c^{\epsilon^{-1}x - n+1} - 1}{c - 1}}([u:v]_{n\epsilon})^{c^{\epsilon^{-1}x - n}}.$$

This completes the proof of theorem 12.1.

Chapter 13

Evolutional dynamics

Let us treat partial differential equations by introducing a time parameter. The idea is also to roughly compare solutions to different PDEs as in the previous chapter, however this extension allows us to compare rather different types of partial differential equations, and can cover much broader classes.

For example later we will compare solutions to the following two partial differential equations:

$$v_s + \epsilon v v_x - \frac{1}{2}v^2 = 0,$$

$$2u_s + \epsilon u(u_s + u_x) = 0.$$

The latter equation contains constants as solutions, but zero is the unique constant solution to the former. One can still say that they share structural similarity in their dynamical framework. This example is obtained by a general procedure which relies on the idea of cancelation by tropical equivalence, which we will also explain later.

13.1 Formal Taylor expansions

A general form of evolutional discrete dynamics is given by:

$$z_{N+1}^{t+1} = f(z_{N-l_0}^{t+1}, \ldots, z_N^{t+1}, z_{N-l_1}^t, \ldots, z_{N+k_1}^t, z_{N-l_2}^{t-1}, \ldots, z_{N+k_2}^{t-1}, \ldots, z_{N+k_{d+1}}^{t-d})$$

where $l_i, k_j \geq 0$ and $N \geq l = \max(l_0, \ldots, l_{d+1})$, $t \geq d$, with initial values:

$$\bar{z}_0^0 \equiv \{ z_a^t \}_{0 \leq a \leq l}^{t=0,1,\cdots} \cup \{ z_N^h \}_{N=0,1,\ldots}^{0 \leq h \leq d}.$$

Let us consider a $C^{\alpha+1}$ function $u : (0, \infty) \times [0, T_0) \rightarrow (0, \infty)$, and introduce other parameters:

$$N = \frac{x}{\epsilon^p}, \quad t = \frac{s}{\epsilon^q}, \quad \epsilon^m u(x, s) = z_N^t$$

where $\epsilon > 0$ is a small constant, and $p, q \geq 1$, $m \geq 0$ are integers. Then we take the Taylor expansions:

$$u(x + i\epsilon^p, s + j\epsilon^q) = u + i\epsilon^p u_x + j\epsilon^q u_s + \frac{(i\epsilon^p)^2}{2} u_{2x} + \frac{(j\epsilon^q)^2}{2} u_{2s}$$

$$+ j\epsilon^q i\epsilon^p u_{xs} + \cdots + \frac{(i\epsilon^p)^\alpha}{\alpha!} u_{\alpha x} + \frac{(j\epsilon^q)^\alpha}{\alpha!} u_{\alpha s}$$

$$+ \frac{(i\epsilon^p)^{(\alpha+1)}}{(\alpha+1)!} u_{(\alpha+1)x}(\xi_{\alpha+1,0}) + \cdots + \frac{(j\epsilon^q)^{(\alpha+1)}}{(\alpha+1)!} u_{(\alpha+1)s}(\xi_{0,\alpha+1})$$

$$\equiv u + i\epsilon^p u_x + j\epsilon^q u_s + \frac{(i\epsilon^p)^2}{2} u_{2x} + \frac{(j\epsilon^q)^2}{2} u_{2s} + j\epsilon^q i\epsilon^p u_{xs}$$

$$+ \cdots + \frac{(i\epsilon^p)^\alpha}{\alpha!} u_{\alpha x} + \frac{(j\epsilon^q)^\alpha}{\alpha!} u_{\alpha s} + \Sigma_{\bar{a}} \frac{(i\epsilon^p)^a (j\epsilon^q)^{\alpha+1-a}}{(\alpha+1)!} u_{\bar{a}}(\xi_{a,\alpha+1-a})$$

where $\bar{a} = (y_{i_1}, \ldots, y_{i_{\alpha+1}})$, $y_j = x$ or s, and $|(x, s) - \xi_{ij}| \leq |(i\epsilon^p, j\epsilon^q)|$.

Let $f = \frac{k}{h} : \mathbb{R}^n_{>0} \to (0, \infty)$ be an elementary rational function, and consider the difference as before:

$$z_{N+1}^{t+1} - f(z_{N-l_0}^{t+1}, \ldots, z_N^{t+1}, z_{N-l_1}^t, \ldots, z_{N+k_1}^t, \ldots, z_{N+k_{d+1}}^{t-d})$$

$$= \epsilon^m u(x + \epsilon^p, s + \epsilon^q)$$

$$- f(\epsilon^m u(x - l_0\epsilon^p, s + \epsilon^q), \ldots, \epsilon^m u(x + k_{d+1}\epsilon^p, s - d\epsilon^q)).$$

By reordering the expansions with respect to the exponents of ϵ, there are rational numbers $a_0, a_1, \cdots \in \mathbb{Q}$ so that the above difference is equal to the following:

$$\epsilon^m \frac{a_0 u + \epsilon^p a_1 u_x + \epsilon^q a_2 u_s + \epsilon^{m+p} a_3 u u_x + \cdots + (i\epsilon^p)^h (j\epsilon^q)^{\alpha+1-h} a_h u_{\bar{h}}(\xi_{\bar{h}}) + \cdots}{h(\epsilon^m u(x - l_0\epsilon^p, s + \epsilon^q), \ldots, \epsilon^m u(x + k_{d+1}\epsilon^p, s - d\epsilon^q))}$$

$$\equiv \frac{1}{h(\epsilon^m u(\ldots))} \times$$

$$\left(\epsilon^m F^1(u) + \epsilon^{m+p} F^2(u_x) + \epsilon^{m+q} F^3(u_s) + \epsilon^{2m+p} F^4(u, u_x) + \cdots \right)$$

where F^k are monomials.

Let us choose finite subsets $A \subset \{1, 2, 3, \ldots\}$, and divide the expanded sum into two terms as:

$$\epsilon^m u(x + \epsilon^p, s + \epsilon^q)$$

$$- f(\epsilon^m u(x - l_0\epsilon^p, s + \epsilon^q), \ldots, \epsilon^m u(x + k_{d+1}\epsilon^p, s - d\epsilon^q))$$

$$= \frac{\epsilon^m F^1(u) + \epsilon^{m+p} F^2(u_x) + \epsilon^{m+q} F^3(u_s) + \epsilon^{2m+p} F^4(u, u_x) + \cdots}{h(\epsilon^m u(x - l_0\epsilon^p, s + \epsilon^q), \ldots, \epsilon^m u(x + k_{d+1}\epsilon^p, s - d\epsilon^q))}$$

$$= \frac{\Sigma_{i \in A} \epsilon^i F^{i'}(u, u_x, u_s, \ldots, u_{\alpha s}) + \Sigma_{j \in A^c} \epsilon^j F^{j'}(u, u_x, \ldots, u_{\bar{a}}(\xi_{\bar{a}}))}{h(\epsilon^m u(x - l_0\epsilon^p, s + \epsilon^q), \ldots, \epsilon^m u(x + k_{d+1}\epsilon^p, s - d\epsilon^q))}$$

$$\equiv \mathbf{F}(\epsilon, u, u_x, u_s, \ldots, u_{\alpha s}) + \epsilon^{m+1} \mathbf{F}^1(\epsilon, u, u_x, \ldots, \{u_{\bar{a}}(\xi_{\bar{a}})\}_{\bar{a}}).$$

As before, we always choose A so that $1 \in A$, and $\mathbf{F}$ does not contain the terms such as $u_{\bar{a}}(\xi)$. We call $\mathbf{F}$ the *leading term* and $\mathbf{F}^1$ the *error term* respectively.

13.2 ϵ-control and higher distortion

Now we return to the starting point. Let f be an elementary rational function, and consider the discrete dynamics:

$$z_{N+1}^{t+1} = f(z_{N-l_0}^{t+1}, \ldots, z_N^{t+1}, z_{N-l_1}^t, \ldots, z_{N+k_{d+1}}^{t-d})$$

as above. One can obtain the leading and error terms $\mathbf{F}$ and $\mathbf{F}^1$ of order α respectively, once the exponents p, q, m of change of variables are determined.

Let the *variational domain* be:

$$D(\epsilon, p, q) = \{(t\epsilon^p a, t\epsilon^q b) : (a, b) \in D, t \in [0, 1]\} \subset \mathbb{R}^2$$

where:

$$D \equiv \{(1, 1), (-l_0, 1), \ldots, (0, 1), (-l_1, 0), \ldots, (k_1, 0),$$
$$(-l_2, -1), \ldots, (k_2, -1), \ldots, (-l_{d+1}, -d), \ldots, (k_{d+1}, -d)\}.$$

For example $D = \{(1, 1), (0, 1), (0, 0), (1, 0), (2, 0), (-1, 1), (0, 1)\}$ for $z_{N+1}^{t+1} = f(z_N^t, z_{N+2}^t, z_{N-1}^{t+1})$.

Regard $\mathbf{F}^1$ as a function on the variables $(x, s, \{\xi_{\bar{a}}\}_{\bar{a}})$. Then we define its ϵ *variation norm*:

$$||\mathbf{F}^1||_\epsilon(x, s) \equiv \sup_{\xi_{\bar{a}} - (x - \epsilon^p, s - \epsilon^q) \in D(\epsilon, p, q)} |\mathbf{F}^1(\epsilon, u(x - \epsilon^p, s - \epsilon^q),$$

$$u_x(x - \epsilon^p, s - \epsilon^q), u_s(x - \epsilon^p, s - \epsilon^q), \ldots, u_{\alpha s}(x - \epsilon^p, s - \epsilon^q), \{u_{\bar{a}}(\xi_{\bar{a}})\}_{\bar{a}})|.$$

Let $u : (0, \infty) \times [0, T_0) \to (0, \infty)$ be a $C^{\alpha+1}$ function.

Definition 13.1. u is ϵ_0 controlled bounded by C, if the ϵ_0 variation norm of $\mathbf{F}^1$ satisfies the pointwise estimates:

$$Cu(x, s) \geq ||\mathbf{F}^1||_{\epsilon_0}(x, s)$$

for all $(x, s) \in (0, \infty) \times [0, T_0)$.

13.2.1 *Admissibility of error term*

Let $u : (0, \infty) \times [0, T_0) \to (0, \infty)$ be a $C^{\alpha+1}$ positive function and f be an elementary rational function. Consider the expansion of the differences:

$$\epsilon^m u(x + \epsilon^p, s + \epsilon^q)$$
$$- f(\epsilon^m u(x - l_0\epsilon^p, s + \epsilon^q), \dots, \epsilon^m u(x + k_{d+1}\epsilon^p, s - d\epsilon^q))$$
$$= \mathbf{F}(\epsilon, u, u_x, u_s, \dots, u_{\alpha s}) + \epsilon^{m+1}\mathbf{F}^1(\epsilon, u, u_x, \dots, \{u_{\bar{a}}(\xi_{\bar{a}})\}_{\bar{a}}).$$

$\mathbf{F}$ has derivative order at most α, while $\mathbf{F}^1$ may contain the terms of the order smaller than $\alpha + 1$ in general.

Let us say that the error term $\mathbf{F}^1$ is *admissible*, if it is of the form:

$$\mathbf{F}^1 = \Sigma_{a \in A^c}\, c_a \epsilon^{s_a} H_a(\epsilon^m u(x - l_0\epsilon^p, s + \epsilon^q), \dots)u_{\bar{a}}(\xi_{\bar{a}})$$

where (1) $|\bar{a}| = \alpha + 1$ and (2) $||H_a(x_1, x_2, \dots)||_{C^0} \leq 1$ for any $x_1, x_2 \cdots \geq 0$.
For this case let the *error constant* be:

$$C_{\mathbf{F}^1} \equiv \Sigma_{a \in A^c}\, |c_a| \in \mathbb{Q}_{\geq 0}.$$

The error constants are determined by the coefficients of rational functions f and of the Taylor expansions.

Let us introduce the *variation norm of u* at order $\alpha + 1$ by:

$$||u||_{\alpha+1}(x, s) = \max_{\partial_i = \partial_x,\, \partial_s}\, \Big\{ \sup_{\xi - (x - \epsilon^p, s - \epsilon^q) \in D(\epsilon, p, q)}\, \Big| \frac{\partial^{\alpha+1} u}{\partial_1 \dots \partial_{\alpha+1}} \Big|(\xi) \Big\}.$$

u is ϵ *controlled at order $\alpha + 1$*, if there is a constant C so that it satisfies the estimates:

$$Cu(x, s) \geq ||u||_{\alpha+1}(x, s)$$

for all $(x, s) \in (0, \infty) \times [0, T_0)$.

Lemma 13.1. *Suppose $\mathbf{F}^1$ is admissible, and u is ϵ controlled at order $\alpha + 1$ bounded by C.*

Then u is ϵ controlled bounded by $CC_{\mathbf{F}^1}$ for some constant $C_{\mathbf{F}^1}$.

Proof. By admissibility, the estimates hold:

$$||\mathbf{F}^1||_\epsilon(x, s) \leq \Sigma_{a \in A^c}|c_a|\epsilon^{s_a}|H_a(\epsilon^m u(x - l_0\epsilon^p, s + \epsilon^q), \dots)|||u||_{\alpha+1}(x, s)$$
$$\leq \Sigma_{a \in A^c}|c_a|||u||_{\alpha+1}(x, s) \leq C_{\mathbf{F}^1}||u||_{\alpha+1}(x, s) \leq C_{\mathbf{F}^1}Cu(x, s).$$

$\square$

Let us introduce the *higher distortion* at order $\alpha + 1$ by:

$$K(u) \equiv \sup_{(x,s) \in (0,\infty) \times [0,T_0]} \frac{||u||_{\alpha+1}(x,s)}{u(x,s)}.$$

In the case when the error term is admissible, finiteness of $K(u)$ is enough to obtain ϵ controlledness.

In the case when the error term is admissible, we obtain the estimate as above:

$$||\mathbf{F}^1||_\epsilon(x,s) \leq C_{\mathbf{F}^1}||u||_{\alpha+1}(x,s) \leq C_{\mathbf{F}^1}K(u)u(x,s).$$

Later we will present examples with admissible error terms.

13.3 Comparison theorem

Let f be an elementary rational function, and consider the evolutional dynamics given by:

$$z_{N+1}^{t+1} = f(z_{N-l_0}^{t+1}, \dots, z_N^{t+1}, z_{N-l_1}^t, \dots, z_{N+k_1}^t, z_{N-l_2}^{t-1}, \dots, z_{N+k_2}^{t-1}, \dots, z_{N+k_{d+1}}^{t-d}).$$

Let g be tropically equivalent to f, and choose the same scaling parameters ϵ, p, q and m. By replacing f by g above, one obtains other leading and error terms $\mathbf{G}$ and $\mathbf{G}^1$ with the equalities:

$$\epsilon^m v(x + \epsilon^p, s + \epsilon^q) -$$
$$g(\epsilon^m v(x - l_0\epsilon^p, s + \epsilon^q), \dots, \epsilon^m v(x + k_{d+1}\epsilon^p, s - d\epsilon^q))$$
$$= \mathbf{G}(\epsilon, v, v_x, v_s, \dots, v_{\alpha s}) + \epsilon^{m+1}\mathbf{G}^1(\epsilon, v, v_x, v_s, \dots, v_{\alpha s}, \{v_{\bar{a}}(\eta_{\bar{a}})\}_{\bar{a}}).$$

Consider the partial differential equations:

$$\mathbf{F}(\epsilon, u, u_x, u_s, \dots, u_{\alpha x}, u_{\alpha s}) = 0,$$
$$\mathbf{G}(\epsilon, v, v_x, v_s, \dots, v_{\alpha x}, v_{\alpha s}) = 0$$

defined for functions $u, v : (0,\infty) \times [0,T_0] \to (0,\infty)$ with positive range.

In order to estimate their ratios $\left(\frac{u(x,s)}{v(x,s)}\right)^{\pm 1}$, we introduce the initial rates.

Definition 13.2. The initial rates are given by:

$$[u : v]_{\epsilon,l,d} \equiv \sup_{(x,s) \in (0,\infty) \times [0,(d+1)\epsilon^q] \cup (0,(l+1)\epsilon^p] \times [0,T_0]} \left(\frac{u(x,s)}{v(x,s)}\right)^{\pm 1}$$

where $l = \max(l_0, \dots, l_{d+1})$.

Recall the initial domain to the corresponding discrete dynamics:

$$\bar{z}_0^0 \equiv \{ z_a^t \}_{0 \leq a \leq l}^{t=0,1,\cdots} \cup \{ z_N^h \}_{N=0,1,\cdots}^{0 \leq h \leq d}.$$

Let c_f and M_f be the Lipschitz constant and the number of the components respectively. Let $c = \max(c_f, c_g)$, $M = \max(M_f, M_g)$, $k = \max(k_1, \ldots, k_{d+1})$ and $L = \max(l, d)$ for $l = \max(l_0, l_1, \ldots, l_{d+1})$.

Theorem 13.1. *Let f and g be elementary rational functions of n variables, which are mutually tropically equivalent. Let $\mathbf{F}$ and $\mathbf{G}$ be their leading terms of order at most $\alpha \geq 0$, and take positive $C^{\alpha+1}$ solutions $u, v : (0, \infty) \times [0, T_0) \to (0, \infty)$ to the equations:*

$$\mathbf{F}(\epsilon, u, u_x, u_s, \ldots, u_{\alpha x}, u_{\alpha s}) = 0,$$

$$\mathbf{G}(\epsilon, v, v_x, v_s, \ldots, v_{\alpha x}, v_{\alpha s}) = 0.$$

Assume both u and v are ϵ_0 controlled bounded by C. Then for any $0 < \epsilon \leq \min(\frac{1}{2C}, \epsilon_0)$ and $D = \max(p, q)$, the estimates hold:

$$\left(\frac{u(x, s)}{v(x, s)}\right)^{\pm 1} \leq (2M)^{8 \frac{c^{\epsilon^{-D}(x+ks)+1} - 1}{c - 1}} ([u : v]_{\epsilon, l, d})^{c^{\epsilon^{-D}(x+ks)}}.$$

The proof is given in sec. 13.4.

Example 13.1. Let $b > a \geq 1$ be positive integers, and consider the discrete dynamics:

$$z_{N+1}^{t+1} = f(z_N^t, z_{N+1}^t) = \frac{1}{b}(a z_N^t + (b - a) z_{N+1}^t).$$

The associated linear partial differential equations are given by $\mathbf{F}(v_x, v_s) = a v_x + b v_s = 0$, by taking the Taylor expansions up to order 2:

$$u(x + \epsilon, s + \epsilon) - f(u(x, s), u(x + \epsilon, s))$$

$$= \frac{\epsilon}{b}(a u_x + b u_s) + \frac{\epsilon^2}{2}((u_{2x} + u_{2s} + 2 u_{xs})(\eta_1) - \frac{b - a}{b} u_{2x}(\eta_2))$$

where we choose the scaling parameters by $N = \frac{x}{\epsilon}$, $t = \frac{s}{\epsilon}$ and $z_N^t = u(x, s)$.

f corresponds to $x_{N+1}^{t+1} = \max(x_N^t, \ldots, x_N^t, x_{N+1}^t, \ldots, x_{N+1}^t) - \max(0, \ldots, 0)$, where their terms iterate a, $b - a$ and b times respectively. Clearly this shows that f are all tropically equivalent independently of $b > a \geq 1$.

Solutions are of the form $u(x, s) = \frac{b}{a} x - s + c$. Choose another g with $b' > a' \geq a$, and compare u with $v(x, s) = \frac{b'}{a'} x - s + c$. Their ratios are uniformly bounded roughly by $2 \max(\frac{ab'}{a'b}, \frac{a'b}{ab'})$ on the domain $[0, \infty) \times [0, \frac{c-1}{a'})$, if $c << (\frac{b}{a})^{\pm 1}, (\frac{b'}{a'})^{\pm 1}$ holds. $\max(\frac{ab'}{a'b}, \frac{a'b}{ab'})$ is bounded by the initial rate.

13.4 Dynamical inequalities

To avoid confusion with another parameter, we will omit the parameter t for f_t and just write f for any relative elementary function.

Let f be an elementary rational function, and consider the evolutional discrete dynamics of the form:

$$z_{N+1}^{t+1} = f(z_{N-l_0}^{t+1}, \ldots, z_N^{t+1}, z_{N-l_1}^t, \ldots, z_{N+k_1}^t, z_{N-l_2}^{t-1}, \ldots, z_{N+k_2}^{t-1}, \ldots, z_{N+k_{d+1}}^{t-d})$$

$$
\begin{array}{ccccc}
z_0^0 & z_1^0 & \cdots & z_N^0 & \cdots \downarrow (t=0) \\
z_0^1 & z_1^1 & \cdots & z_N^1 & \cdots \downarrow (t=1) \\
\vdots & \vdots & \vdots & \vdots & \vdots \\
z_0^t & z_1^t & \cdots & z_N^t & \cdots \\
\vdots & \vdots & \vdots & \vdots & \vdots
\end{array}
$$

Let g be tropically equivalent to f, and consider the orbit $\{w_n^t\}$ by g with any initial value $\bar{w}_0^0$.

Introduce the *initial rate* as:

$$[\bar{z}_0^0 : \bar{w}_0^0] = \sup \left\{ \; \left\{ \frac{z_a^t}{w_a^t} \right\}_{0 \le a \le l}^{t=0,1,\ldots}, \quad \left\{ \frac{z_N^h}{w_N^h} \right\}_{N=0,1,\ldots}^{0 \le h \le d} \; \right\}.$$

Let us put $l = \max(l_0, l_1, \ldots, l_{d+1})$, $k = \max(k_1, \ldots, k_{d+1})$ and

$$A(N, t) \equiv (t - d - 1)k + N - l$$

for $N \ge l + 1$ and $t \ge d + 1$.

Proposition 13.1. *(1) Let f and g be tropically equivalent, and consider their orbits $\{z_N^t\}_N$ and $\{w_N^t\}_N$ for f and g with the initial values $\bar{z}_0^0$ and $\bar{w}_0^0$ respectively. Then they satisfy the estimates:*

$$\left(\frac{z_N^t}{w_N^t}\right)^{\pm} \le M^{4P_{A(N,t)}(c)} [\bar{z}_0^0 : \bar{w}_0^0]^{c^{A(N,t)}}$$

where $c = \max(c_f, c_g)$ and $M = \max(M_f, M_g)$ and $P_i(c) = \frac{c^{i+1}-1}{c-1}$.

(2) Let f, f', g, g' be four elementary rational functions which are all tropically equivalent. Let $\{u_N^t\}_{N,t}$ and $\{v_N^t\}_{N,t}$ be positive sequences so that these satisfy the estimates:

$$f'(u_{N-l_0}^{t+1}, \ldots, u_{N+k_{d+1}}^{t-d}) \le u_{N+1}^{t+1} \le f(u_{N-l_0}^{t+1}, \ldots, u_{N+k_{d+1}}^{t-d}),$$
$$g'(v_{N-l_0}^{t+1}, \ldots, v_{N+k_{d+1}}^{t-d}) \le v_{N+1}^{t+1} \le g(v_{N-l_0}^{t+1}, \ldots, v_{N+k_{d+1}}^{t-d})$$

for $N \geq l$ and $t \geq d$. Then the ratios satisfy the uniform estimates:

$$\left(\frac{v_N^t}{u_N^t}\right)^{\pm 1} \leq M^{8P_{A(N,t)}(c)}[\bar{u}_0^0 : \bar{v}_0^0]c^{A(N,t)}$$

for $N \geq l+1$ and $t \geq d+1$.

Proof. Let us check that in order to determine z_{l+N}^{d+t}, one has to iterate at most $(t-1)k + N$ times to apply f for $N, t \geq 1$. Then the conclusions follow from theorem 3.1 and proposition 12.1 (see remark 3.1).

Let us denote by $\alpha(N, t)$ the number of compositions of f in order to determine z_N^t. It is an increasing function on both variables. We show the estimates $\alpha(l + N, d + t) \leq (t-1)k + N$.

Let $\Delta_0 = \{(a,b) \in \{0, 1, \ldots, k+l\} \times \{0, 1, \ldots, d\} \cup \{0, \ldots, l\} \times \{d+1\}\}$ be the finite set. This is a basic building block in the sense that for $N, t \geq 1$, z_{N+l}^{t+d} is determined if one knows z_{N-1+a}^{t-1+b} for $(a,b) \in \Delta_0$.

We proceed by induction on t. $\alpha(l + N, d + 1) \leq N$ clearly follows.

Suppose the conclusion follows for $t \leq t_0$, and so $\alpha(N + l, d + t_0) \leq (t_0 - 1)k + N$ holds. Then:

$$\begin{aligned}
\alpha(l+1, d+t_0+1) &= \alpha(l+k, d+t_0) + 1 \\
&\leq (t_0 - 1)k + k + 1 = t_0 k + 1
\end{aligned}$$

holds. Next:

$$\begin{aligned}
\alpha(l+2, d+t_0+1) &= \max(\alpha(l+1, d+t_0+1), \alpha(l+k+1, d+t_0)) + 1 \\
&\leq t_0 k + 2.
\end{aligned}$$

Using the estimates:

$$\alpha(N+l, d+t_0+1) \leq \max(\alpha(N-1+l, d+t_0+1), \alpha(N-1+l+k, d+t_0)) + 1$$

one can obtain the bounds $\alpha(N + l, d + t_0 + 1) \leq t_0 k + N$.

$\square$

Proof of theorem 13.1. Recall $A(N, t) = (t - d - 1)k + N - l$ for $N \geq l+1$ and $t \geq d+1$. Take $0 < \epsilon \leq \min(\frac{1}{2C}, \epsilon_0)$. Combining with proposition 13.1, the parallel argument to the proof of theorem 12.1 gives the estimates:

$$\left(\frac{u(N\epsilon^p + \mu, t\epsilon^q + \chi)}{v(N\epsilon^p + \mu, t\epsilon^q + \chi)}\right)^{\pm 1} \leq (2M)^{8P_{A(N,t)}(c)}([u : v]_{\epsilon,l,d})^{c^{A(N,t)}}$$

for any $0 \le \mu \le \epsilon^p$ and $0 \le \chi \le \epsilon^q$. Then we have the estimates:

$$A(N,t) = (t - d - 1)k + N - l$$
$$\le \epsilon^{-q} k(t\epsilon^q + \chi) - dk + \epsilon^{-p}(N\epsilon^p + \mu) - l$$
$$\le \epsilon^{-q} k(t\epsilon^q + \chi) + \epsilon^{-p}(N\epsilon^p + \mu)$$
$$\le \epsilon^{-D}[k(t\epsilon^q + \chi) + (N\epsilon^p + \mu)]$$

where $D = \max(p, q)$. Then:

$$(2M)^{8P_{A(N,t)}(c)} [u : v]^{c^{A(N,t)}}_{\epsilon,l,d}$$
$$\le (2M)^{8 \frac{c^{\epsilon^{-D}[k(t\epsilon^q+\chi)+(N\epsilon^p+\mu)]+1}-1}{c-1}} ([u : v]_{\epsilon,l,d})^{c^{\epsilon^{-D}[k(t\epsilon^q+\chi)+(N\epsilon^p+\mu)]}}.$$

Now combining these estimates, one obtains the desired estimates:

$$\left(\frac{u(x,s)}{v(x,s)} \right)^{\pm 1} \le (2M)^{8 \frac{c^{\epsilon^{-D}(x+ks)+1}-1}{c-1}} ([u : v]_{\epsilon,l,d})^{c^{\epsilon^{-D}(x+ks)}}.$$

This completes the proof of theorem 13.1.

13.5   Quasi-linear equations

Below we apply the general procedure above to quasi-linear equations of order 1. Our strategy is as follows: given a partial differential equation, find a 'good' elementary rational function f.

Here we introduce a cancelation method of non linear terms, and apply it to compare solutions between the following equations:

$$v_s + \epsilon v v_x - \frac{1}{2} v^2 = 0,$$
$$2u_s + \epsilon u(u_s + u_x) = 0$$

where $\epsilon > 0$ is a small constant. The former equation admits a degenerate solution $v(x,s) = \frac{c}{1-0.5cs}$ on $(0, \infty) \times [0, \frac{2}{c})$ for $c > 0$.

We choose the variation norm at order 2:

$$\|u\|_2(x,s) = \sup_{\xi - (x-\epsilon, s-\epsilon) \in D(\epsilon,1,1)} \left\{ \left| \frac{\partial^2 u}{\partial x^2} \right|(\xi), \left| \frac{\partial^2 u}{\partial s^2} \right|(\xi), \left| \frac{\partial^2 u}{\partial x \partial s} \right|(\xi) \right\}$$

and put the higher distortion at order 2:

$$K(u) \equiv \sup_{(x,s) \in (0,\infty) \times [0,T_0)} \frac{\|u\|_2(x,s)}{u(x,s)}.$$

Let us fix any positive constant $K_0 > 0$.

Theorem 13.2. *For any* $0 < \epsilon \le 0.1K_0^{-1}$, *let* $v, u : (0, \infty) \times [0, T_0) \to (0, \infty)$ *be* C^2 *solutions to the quasi-linear equations:*

$$v_s + \epsilon v v_x - \frac{1}{2}v^2 = 0,$$

$$2u_s + \epsilon u(u_s + u_x) = 0.$$

Suppose their higher distortions $K(u), K(v)$ *at order* 2 *are bounded by* K_0. *Then they satisfy the asymptotic estimates for all* $(x, s) \in (0, \infty) \times [0, T_0)$:

$$\left(\frac{u(x, s)}{v(x, s)}\right)^{\pm 1} \le 40^{2^{\epsilon^{-1}(x+2s)+1}} \left([u : v]_{\epsilon, 1, 0}\right)^{2^{\epsilon^{-1}(x+2s)}}.$$

In particular when $u(x, s) \equiv R > 0$ *is constant, then the estimates hold:*

$$R(40)^{-2^{\epsilon^{-1}(x+2s)+1}}\left([v : R]_{\epsilon, 1, 0}\right)^{-2^{\epsilon^{-1}(x+2s)}}$$

$$\le v(x, s) \le R(40)^{2^{\epsilon^{-1}(x+2s)+1}}\left([v : R]_{\epsilon, 1, 0}\right)^{2^{\epsilon^{-1}(x+2s)}}.$$

13.6   Induction of the equations

Let us start from a concrete case, and consider the discrete dynamics:

$$z_{N+1}^{t+1} = f(z_N^t, z_{N+2}^t, z_{N-1}^{t+1}) \equiv \frac{z_{N+2}^t}{2} + \frac{z_N^t(1 + 2z_{N-1}^{t+1})}{2(1 + z_N^t)}.$$

The corresponding $(\max, +)$-function is given by:

$$\max(x_{N+2}^t, x_{N+2}^t + x_N^t, x_N^t, x_N^t + x_{N-1}^{t+1}, x_N^t + x_{N-1}^{t+1}) - \max(0, 0, x_N^t, x_N^t).$$

The number of the components is $M = 5 \times 4 = 20$, and its Lipschitz constant is equal to 2.

We choose the scaling parameters by:

$$\epsilon v(x, s) = z_N^t, \quad N = \frac{x}{\epsilon}, \quad t = \frac{s}{\epsilon}$$

where we take a small $\epsilon > 0$ so that the estimate $\epsilon \le 0.1V_0^{-1}$ holds.

Let $v : (0, \infty) \times [0, T_0) \to (0, \infty)$ be a C^2 function, and take the Taylor expansions up to order 2:

$v(x+i\epsilon, s + j\epsilon)$

$$= v + i\epsilon v_x + j\epsilon v_s + \epsilon^2\left(\frac{i^2}{2}v_{2x}(\xi_{2i,0}) + \frac{j^2}{2}v_{2s}(\xi_{0,2j}) + ijv_{xs}(\xi_{i,j})\right)$$

$$\equiv v + i\epsilon v_x + j\epsilon v_s + \epsilon^2 D^2 v(\xi_{ij}).$$

Let us insert the formal Taylor expansions:

$$\epsilon v(x + \epsilon, s + \epsilon) - f(\epsilon v(x, s), \epsilon v(x + 2\epsilon, s), \epsilon v(x - \epsilon, s + \epsilon))$$

$$= \frac{1}{2(1 + \epsilon v)}[\epsilon^2(2v_s + 2\epsilon v v_x - v^2) - 2(\epsilon v)\epsilon^2 D^2 v(\xi_{-11})\}]$$

$$+ \epsilon^2(D^2 v(\xi_{11}) - \frac{1}{2}D^2 v(\xi_{20}))$$

where the leading term is given by:

$$\mathbf{F} = \epsilon^2 \frac{2v_s + 2\epsilon v v_x - v^2}{2(1 + \epsilon v)}$$

which is computed from the equality:

$$2\epsilon(1 + \epsilon v)(v + v_s + v_x) = \epsilon(1 + \epsilon v)(v + 2v_x) + \epsilon v + 2\epsilon^2 v(v + v_s - v_x).$$

The error term is admissible, so let us calculate the error constant $C_{\mathbf{F1}}$. Notice that the estimates $|D^2 v(\xi_{ij})| \leq (\frac{i^2 + j^2}{2} + |ij|)||v||_2(x, s)$, where $||v||_2(x, s)$ is the variation norm at order 2. Then the error term satisfies the estimates:

$$||\mathbf{F}^1||_\epsilon(x, s) \leq \frac{2(\epsilon v)\epsilon^2 |D^2 v(\xi_{-11})|}{2(1 + \epsilon v)} + \epsilon^2(|D^2 v(\xi_{11})| + |\frac{1}{2}D^2 v(\xi_{20})|)$$

$$\leq 5\epsilon^2 ||v||_2(x, s).$$

In particular the error constant is given by $C_{\mathbf{F1}} = 5$.

13.7 Deformation and cancelation

Let us introduce a cancelation method, which is explained by the example below.

Let us consider the discrete dynamics:

$$w_{N+1}^{t+1} = g(w_N^t, w_{N+2}^t, w_{N-1}^{t+1}) \equiv \frac{w_{N+2}^t}{2} + \frac{w_N^t(1 + w_{N-1}^{t+1})}{2(1 + w_N^t)}.$$

g is tropically equivalent to f. The number of the components is $4 \times 4 = 16$, and the corresponding $(\max, +)$-function has its Lipschitz constant 2.

Let $u : (0, \infty) \times [0, T_0) \to (0, \infty)$ be a C^2 function, and choose the same scaling parameters, $\epsilon u(x, s) = z_N^t$, $N = \frac{x}{\epsilon}$ and $t = \frac{s}{\epsilon}$. Then insert the Taylor expansions of u up to order 2 into the difference as before. The direct calculation shows that unlike the previous case, the u^2 term is eliminated, and the result is given by:

$$\epsilon u(x + \epsilon, s + \epsilon) - g(\epsilon u(x, s), \epsilon u(x + 2\epsilon, s), \epsilon u(x - \epsilon, s + \epsilon))$$

$$= \frac{1}{2(1 + \epsilon u)}[\epsilon^2(2u_s + \epsilon u u_s + \epsilon u u_x) - \epsilon u \epsilon^2 D^2 u(\eta_{-11})]$$

$$+ \epsilon^2(D^2 u(\eta_{11}) - \frac{1}{2}D^2 u(\eta_{20}))$$

where the leading term is given by:
$$\mathbf{G} = \epsilon^2 \frac{2u_s + \epsilon u u_s + \epsilon u u_x}{2(1 + \epsilon u)}$$
which is computed from the equality:
$$2\epsilon(1 + \epsilon u)(u + u_s + u_x) = \epsilon(1 + \epsilon u)(u + 2u_x) + \epsilon u + \epsilon^2 u(u + u_s - u_x).$$

In this deformation also, the error term is admissible, and satisfies the estimates:
$$\|\mathbf{G}^1\|_\epsilon(x, s) \le \frac{\epsilon u \epsilon^2 |D^2 u(\eta_{-11})|}{2(1 + \epsilon u)} + \epsilon^2(|D^2 u(\eta_{11})| + |\frac{1}{2} D^2 u(\eta_{20})|)$$

$$\le 4\epsilon^2 \|u\|_2(x, s).$$
So the error constant is give by $C_{\mathbf{G}^1} = 4$.

Proof of theorem 13.2. Let $u, v : (0, \infty) \times [0, T_0) \to (0, \infty)$ be C^2 functions which satisfy the equations $v_s + \epsilon v v_x - \frac{1}{2} v^2 = 0$ and $2u_s + \epsilon u(u_s + u_x) = 0$.

Suppose they have bounded higher distortion $K(u), K(v) \le K_0$ at order 2. Then by applying theorem 13.1 and lemma 13.1, one obtains the asymptotic estimates:
$$\left(\frac{u(x, s)}{v(x, s)}\right)^{\pm 1} \le (2M)^{8 \frac{c^{\epsilon - D}(x+ks)+1_{-1}}{c-1}} ([u : v]_{\epsilon, 1, 0})^{c^{\epsilon - D}(x+ks)+n}$$
for any $0 < \epsilon \le (2CV_0)^{-1}$, where in this case $D = \max(p, q) = 1$, $C = 5$, $L = 1$, $M = 20$, $c = 2$ and $k = 2$. Thus for any $0 < \epsilon \le 0.1V_0^{-1}$, the estimates:
$$\left(\frac{u(x, s)}{v(x, s)}\right)^{\pm 1} \le (40)^{2^{\epsilon - 1}(x+2s)+1} ([u : v]_{\epsilon, 1, 0})^{2^{\epsilon - 1}(x+2s)}$$
hold. This completes the proof of theorem 13.2.

Remark 13.1. This is an instance of asymptotic comparison between solutions to different partial differential equations by applying the cancelation method. One may try for other cases using various elementary rational functions and their tropical equivalent functions. See also [Kat8] for other cases of PDEs.

References

Chapters 12 and 13 gave a comparison analysis of the induced partial differential equations in [Kat8], which include some concrete examples. For basic analysis on quasi-linear equations, see [Log].

Chapter 14

Rough analytic relation on the set of partial differential equations

Tropical geometry provides us with an important prototype in mathematical physics. Particularly we have seen that the Korteweg–de Vries (KdV) equation admits BBS automaton as its framework. It allows us to study analytic aspects at the same time for three categories of dynamical systems which sit in different hierarchies mutually. The classes of dynamics which can be analyzed by these scaling limits are rather broad beyond integrable systems. In fact it is easy to find discrete dynamical systems which are far from integrable, but which are transformed into integrable cell automata. If we consider a situation when two dynamical systems are transformed into the same integrable cell automaton, where one is integrable and the other is not, then it will be quite natural to try to classify such dynamical systems by some systematic method, which should include wide classes of dynamical systems as above.

Recall that in the previous chapter, we have constructed the comparison method to give rough asymptotic estimates for solutions which obey different partial differential equations. Roughly speaking, two different solutions can be compared when their defining equations are transformed to the same automata at infinity.

Based on this method, we introduce analytic relations on the set of partial differential equations of two variables. Our construction presents a systematic way to give related pairs of different partial differential equations. We also construct some unrelated pairs concretely. These constructions lead to the fact that the new relations are non trivial. We also make numerical calculations and compare the results for both related and unrelated pairs of PDEs.

14.1   Summary

Let:
$$\mathbf{PDE}_2 = \{P(\epsilon, u, u_x, u_s, u_{xs}, \ldots, u_{\alpha s}) = 0 :$$
$$P \text{ are polynomials with real coefficients}\}$$
be the set of all families of polynomial type partial differential equations with 2 variables parametrized by ϵ.

We shall introduce some analytic relations $\sim$ among elements in $\mathbf{PDE}_2$, which are characterized by uniform asymptotic estimates for all positive solutions among different PDEs. Actually there are four types of the relations with respect to growth rates, exponential or double exponential of solutions, and to bounded or unbounded domains of solutions on the space variable.

Let us describe what can be said roughly. We will state them precisely in theorem 14.1 and proposition 14.1 later.

The following result is a reformulation of theorem 13.1:

'Theorem' 14.1. *Suppose two partial differential equations P, Q are obtained from two elementary rational functions f and g by the process in chapter 13. Assume moreover that they are mutually tropically equivalent. Then they are mutually related.*

As a concrete case, two partial differential equations of order 1:
$$v_s + \frac{\epsilon}{2} v v_x - \frac{1}{2} v^2 = 0,$$
$$2u_s + \frac{\epsilon}{2} u(u_s + u_x) = 0$$
are mutually related by theorem 13.2.

Let us describe the unrelated case roughly.

'Proposition' 14.1. (1) *The following two linear partial differential equations of order l:*
$$\epsilon u_s + \frac{\epsilon^2}{2} u_{2s} + \cdots + \frac{\epsilon^l}{l!} u_{ls} + \epsilon^2 u_{xs} + \frac{\epsilon^3}{2} u_{xss} + \cdots + \frac{\epsilon^l}{l!} u_{(l-1)M's} = 0,$$
$$\epsilon(v_s + v_x) + \epsilon^2 (v_{xs} + \frac{1}{2} v_{2s} + \frac{1}{2} v_{2x}) + \cdots + \frac{\epsilon^l}{l!} v_{lx} = 0$$
are mutually unrelated.

(2) *The following two linear partial differential equations:*
$$4u_s + \epsilon u(u_s + u_x) = 0,$$
$$v_s = I v_x$$

are mutually unrelated.

To obtain unrelated pairs, one has to seek their solutions which break uniform bounds which appear in theorem 13.1. The analysis for (2) touches on the technique in the field of the *singular perturbations* in the sense of continuity of solutions at $\epsilon = 0$.

We also include computer calculations which seem quite effective to understanding the behavior of solutions. We will see that certainly their data reflect related and unrelated situations.

Remark 14.1. (1) These two opposite statements imply that the relations are non trivial. It would be of particular interest for us to measure the 'sizes' of the related partial differential equations of 2 variables.

(2) In order to treat real models in physics, we need to extend this method for systems of partial differential equations. In particular KdV equation is quite an intriguing system and has surprising relationship with the soliton cellular automaton system known as box and ball system. We would expect that study in such a direction might lead us to rough classifications of non equilibrium systems.

14.2 Set of partial differential equations

Recall the contents in chapter 13. Our basic idea is to approximate PDE by discrete rational dynamics of the form:

$$z_{N+1}^{t+1} = f(z_{N-l_0}^{t+1}, \ldots, z_N^{t+1}, z_{N-l_1}^t, \ldots, z_{N+k_1}^t, \ldots, z_{N+k_{d+1}}^{t-d})$$

by introducing scaling parameters $z_N^t = \epsilon^m u(x, s)$ and $(N, t) = (\epsilon^{-p} x, \epsilon^{-q} s)$. Notice that for any n variable function f as above, one can choose various types of the sets of variables $(z_{N-l_0}^{t+1}, \ldots, z_N^{t+1}, z_{N-l_1}^t, \ldots, z_{N+k_1}^t, \ldots, z_{N+k_{d+1}}^{t-d})$. We say that the sets:

$$\{(z_{N-l_0}^{t+1}, \ldots, z_N^{t+1}, z_{N-l_1}^t, \ldots, z_{N+k_1}^t, \ldots, z_{N+k_{d+1}}^{t-d}), (m, p, q)\}$$

are the *approximation data*.

We say that the above rational dynamics is *consistent*, if the following bounds hold:

$$k_1, \ldots, k_{d+1} \leq 1.$$

As a general procedure in 13.1, the rational dynamics with the scaling parameters above give a parametrized partial differential equation

$\mathbf{F}(\epsilon, u, u_x, \dots) = 0$ as the leading term, and the error term $\mathbf{F}^1(\epsilon, u, u_x, \dots)$ using Taylor expansions:

$$z_{N+1}^{t+1} - f(z_{N-l_0}^{t+1}, \dots, z_{N+k_{d+1}}^{t-d}) = \epsilon^m \mathbf{F}(\epsilon, u, u_x, u_s, u_{xs}, \dots, u_{\alpha x}, u_{\alpha s})$$
$$+ \epsilon^{m+1} \mathbf{F}^1(\epsilon, u, u_x, u_s, u_{xs}, \dots, u_{\alpha s}, u_{(\alpha+1)x}(\xi_{\alpha+1,0}), \dots, u_{(\alpha+1)s}(\xi_{0,\alpha+1})).$$

Definition 14.1. A partial differential equation:

$$P(u, u_x, u_s, u_{xs}, \dots, u_{\alpha x}, u_{\alpha s})$$

is in $\mathbf{PDE}_2^\infty$, if there is an induced pair $(\mathbf{F}, \mathbf{F}^1)$ as above and a positive function $h > 0$ so that:

(1) $\mathbf{F}$ satisfies the equality:

$$\mathbf{F}(\epsilon, u, u_x, u_s, u_{xs}, \dots, u_{\alpha x}, u_{\alpha s}) = \frac{P(u, u_x, u_s, u_{xs}, \dots, u_{\alpha x}, u_{\alpha s})}{h(u, u_x, u_s, u_{xs}, \dots, u_{\alpha x}, u_{\alpha s})}$$

(2) there is a constant $C \geq 0$ so that the pointwise estimates hold:

$$|\mathbf{F}^1(\epsilon, u, u_x, u_s, \dots, u_{\alpha s}, u_{(\alpha+1)x}(\xi_{\alpha+1,0}), u_{\alpha x,s}(\xi_{\alpha,1}), \dots, u_{(\alpha+1)s}(\xi_{0,\alpha+1}))|$$
$$\leq C(|u_{(\alpha+1)x}(\xi_{\alpha+1,0})| + |u_{\alpha x,s}(\xi_{\alpha,1})| + \dots + |u_{(\alpha+1)s}(\xi_{0,\alpha+1})|).$$

P is said to be in $\mathbf{PDE}_2^{fin}$, if in addition the discrete dynamics is consistent. The domains of solutions are given by:

$$(0, A_0) \times [0, T_0] \quad \text{if } P \in \mathbf{PDE}_2^{fin}, \quad A_0, T_0 \in (0, \infty],$$
$$(0, \infty) \times [0, T_0] \quad \text{if } P \in \mathbf{PDE}_2^\infty, \quad T_0 \in (0, \infty].$$

Let $||u||_{\alpha+1}$ be the C^0 norm of the variation norm of u at order $\alpha + 1$:

$$||u||_{\alpha+1} = \max_{\partial_i = \partial_x, \partial_s} \left\{ \left|\left| \frac{\partial^{\alpha+1} u}{\partial_1 \dots \partial_{\alpha+1}} \right|\right|_{C^0((0,A_0) \times [0,T_0))} \right\}.$$

Then we obtain an *a priori* estimate of the last term:

$$C(|u_{(\alpha+1)x}(\xi_{\alpha+1,0})| + \dots + |u_{(\alpha+1)s}(\xi_{0,\alpha+1})|) \leq CL \, ||u||_{\alpha+1}$$

where L is the number of the summation of $\alpha + 1$ derivatives of u. We say that the number CL is the *error constant* for the approximation of P. Notice that the error constant is determined by the approximation data and the original rational function f.

14.3 Classes of partial differential equations

Let $u : (0, A_0) \times [0, T_0) \to (0, \infty)$ be in $C^{\alpha+1}$ where $A_0, T_0 \in (0, \infty]$. Recall two data in 12.2:

(1) The *initial rate*:
$$[u : v]_\epsilon \equiv \sup_{(x,s) \in (0,A_0) \times [0,\epsilon] \cup (0,\epsilon] \times [0,T_0)} \left(\frac{u(x, s)}{v(x, s)} \right)^{\pm 1}.$$

(2) The *higher distortion* of order $\alpha + 1$ by:
$$K_{\alpha+1}(u) \equiv \frac{\|u\|_{\alpha+1}}{\inf_{(x,s) \in (0,A_0) \times [0,T_0)} u(x, s)}.$$

Let us introduce analytic relations on the set of partial differential equations. Denote:
$$(*, +) \in \{(fin, e), (fin, e^e), (\infty, e), (\infty, e^e)\}.$$

Definition 14.2. Let $P, Q \in \mathbf{PDE}_2^*$ of order α. P and Q are related in $(*, +)$:
$$P \sim_*^+ Q$$
if there are constants $M, c, D, L \geq 1$ and C, C' so that for any positive solution $u, v : (0, A_0) \times [0, T_0) \to (0, \infty)$:
$$P(\epsilon, u, u_x, u_s, u_{xs}, \dots, u_{\alpha s}) = 0,$$
$$Q(\epsilon, v, v_x, v_s, v_{xs}, \dots, v_{\alpha s}) = 0$$
they satisfy the asymptotic estimates:
$$\left(\frac{u(x, s)}{v(x, s)} \right)^{\pm 1} \leq \begin{cases} M^{c^{\epsilon-D}(x+s+1)} ([u : v]_{L\epsilon})^{c^{\epsilon-D}(x+s+1)} & + = e^e, \\ M^{\epsilon-D}(x+s+1) [u : v]_{L\epsilon} & + = e \end{cases}$$
for all $0 < L\epsilon \leq \min(\frac{1}{CK}, A_0, T_0, C')$, where $K = \max(K_{\alpha+1}(u), K_{\alpha+1}(v))$.

Remark 14.2. Thus there are four cases of the relations in total.

Let $P, Q \in \mathbf{PDE}_2^{fin}$. $P_1 \sim_\infty^+ P_2$ holds whenever $P_1 \sim_{fin}^+ P_2$ for both $+ = e$ or e^e. The other cases are described by the diagram below:

$$\sim_{fin}^e \ \leq \ \sim_{fin}^{e^e}$$

$$|\wedge \qquad |\wedge$$

$$\sim_\infty^e \ \leq \ \sim_\infty^{e^e}$$

where $\sim_*^+ \ \leq \ \sim_{*'}^{+'}$ means that $u \sim_*^+ v$ implies $u \sim_{*'}^{+'} v$ for any two solutions.

When we specify the data, we denote:

$$P \sim_*^{e^e} Q \quad \text{in } (M, c, D, L; \alpha), \quad \text{or } P \sim_*^e Q \quad \text{in } (M, 1, D, L; \alpha).$$

The following is immediate:

Lemma 14.1. *Suppose $M \le M'$, $c \le c'$, $D \le D'$ and $L \le L'$ hold. Then $P_1 \sim_*^+ P_2$ in $(M, c, D, L; \alpha)$ implies $P_1 \sim_*^+ P_2$ in $(M', c', D', L'; \alpha)$.*

Remark 14.3. Notice that even though two pairs $P_1, P_2 \in \mathbf{PDE}_2^*$ are unrelated in $\sim_*^+$ in the class $(M, c, D, L; \alpha)$, they might still be related in $\sim_*^+$ in the class $(M', c', D', L'; \alpha)$.

For $P \in \mathbf{PDE}_2^{fin}$, we consider $(0, A_0) \times [0, T_0)$ as domains of solutions, and $(0, \infty) \times [0, T_0)$ for $P \in \mathbf{PDE}_2^\infty$, where $A_0, T_0 \in (0, \infty]$.

Let:

$$\{(z_{N-l_0}^{t+1}, \ldots, z_N^{t+1}, z_{N-l_1}^t, \ldots, z_{N+k_1}^t, \ldots, z_{N+k_{d+1}}^{t-d}), (m, p, q)\}$$

be an approximation data. It gives the following data:

$$L = \max(l, d), \ D = \max(p, q), \ k = \max(k_1, \ldots, k_{d+1}), \ l = \max(l_0, \ldots, l_{d+1}).$$

For an elementary rational function f, we have the number of the components and the Lipschitz constant $M = M_f$ and $c = c_f$ respectively. So an approximation data and f present these data:

$$(M, c, D, L, k).$$

We say that $P(u, u_x, u_s, u_{xs}, \ldots, u_{\alpha s}) \in \mathbf{PDE}_2^*$ is *approximable* from f in (M, c, L, k, D), if it is induced from discrete dynamics by f as above whose constants are all less than or equal to M, c, L, k, D respectively.

Remark 14.4. Notice that three data, defining equations of discrete dynamics (1) $z_{N+1}^{t+1} = f(z_{N-l_0}^{t+1}, \ldots, z_{N+k_{d+1}}^{t-d})$, (2) the exponents of the scaling change of variables (m, p, q) and (3) the order to take the Taylor expansions α, determine the defining partial differential equations.

14.4 Asymptotic estimates

Let us choose an approximation data and let L, k, D be the corresponding data. Let f and g be two elementary rational functions of n variables. We have the extra data:

$$M = \max(M_f, M_g), \quad c = \max(c_f, c_g), \quad C$$

where C is the maximum of the error constants on their error terms.

Recall theorem 13.1:

Theorem 14.1. *Suppose f and g are mutually tropically equivalent.*

For $ = $ fin or ∞, let $P, Q \in \mathbf{PDE}_2^*$ be two partial differential equations of order $\alpha \geq 0$ as their leading terms corresponding to f and g respectively.*

Let us take positive $C^{\alpha+1}$ solutions $u, v : (0, A_0) \times [0, T_0) \to (0, \infty)$ with:

$$P(\epsilon, u, u_x, u_s, \ldots, u_{\alpha x}, u_{\alpha s}) = 0,$$
$$Q(\epsilon, v, v_x, v_s, \ldots, v_{\alpha x}, v_{\alpha s}) = 0$$

and assume the estimates $K_{\alpha+1}(u), K_{\alpha+1}(v) \leq K$.

Then for any $0 < \epsilon \leq \min(\frac{1}{2CK}, (L+1)^{-1}A_0, (L+1)^{-1}T_0, n^{-1})$, they satisfy the asymptotic estimates:

$$\left(\frac{u(x,s)}{v(x,s)}\right)^{\pm 1} \leq (2M)^{8\frac{c^{\epsilon^{-D}(x+ks)+1}-1}{c-1}} ([u:v]_{(L+1)\epsilon})^{c^{\epsilon^{-D}(x+ks)}}$$

where the constants are the approximation data above.

If $c \leq 1$, then they admit the exponential asymptotic estimates.

Corollary 14.1. *Under the above situation, $P_1 \sim_*^+ P_2$ in $(M', c', D, L+1; \alpha)$, where:*

$$(M', c') = \begin{cases} ((2M)^{\frac{8}{c-1}}, c^{k'}) & c > 1, \\ ((2M)^{8k'}, 1) & c = 1 \end{cases}$$

and $k' = \max(1, k)$.

Proof. Firstly suppose $c > 1$. Then one has the estimates:

$$(2M)^{8\frac{c^{\epsilon^{-D}(x+ks)+1}-1}{c-1}} \leq [(2M)^{\frac{8}{c-1}}](c^{k'})^{\epsilon^{-D}(x+s+1)},$$
$$c^{\epsilon^{-D}(x+ks)} \leq (c^{k'})^{\epsilon^{-D}(x+s+1)}.$$

Next suppose $c = 1$. Then:

$$\lim_{c \to 1}(2M)^{8\frac{c^{\epsilon^{-D}(x+ks)+1}-1}{c-1}} = (2M)^{8(\epsilon^{-D}(x+ks)+1)} \leq [(2M)^{8k'}]^{\epsilon^{-D}(x+s+1)}.$$

$\square$

Lemma 14.2. *Two partial differential equations of order 1:*

$$v_s + \frac{\epsilon}{2}vv_x - \frac{1}{2}v^2 = 0,$$

$$2u_s + \frac{\epsilon}{2}u(u_s + u_x) = 0$$

are both in $\mathbf{PDE}_2^\infty$ *which are approximable by some elementary rational functions in* $(20, 2, 1, 2, 1)$, *and they are mutually related in* $\sim_\infty^{e^e}$ *in the class* $(40, 4, 1, 2; 1)$

Proof. Recall theorem 13.2. We have verified that the above equations are in $\mathbf{PDE}_2^\infty$ and their error constants are both bounded by $5\epsilon \leq 5$.

Moreover any C^2 positive solutions satisfy the asymptotic estimates for all $(x, s) \in (0, \infty) \times [0, T_0)$:

$$\left(\frac{u(x, s)}{v(x, s)}\right)^{\pm 1} \leq 40^{2^{\epsilon^{-1}(x+2s)+1}}([u : v]_{2\epsilon})^{2^{\epsilon^{-1}(x+2s)}}$$

$$\leq 40^{2^{2\epsilon^{-1}(x+s+1)}}([u : v]_{2\epsilon})^{2^{2\epsilon^{-1}(x+s+1)}}$$

where $K = \max(K_2(u), K_2(v))$ and $0 < \epsilon \leq \min(0.1K^{-1}, 1/3)$. $\qquad\square$

14.5 Unrelated classes

The related pairs we have presented essentially arose from the construction of asymptotic comparisons in chapter 13. To show our relations are non trivial, we have to present the existence of unrelated pairs.

Let us consider discrete dynamics $z_{N+1}^{t+1} = f(z_{N-l_0}^{t+1}, \ldots, z_{N+k_{d+1}}^{t-d})$ and fix (m, p, q). Then one obtains a family of partial differential equations $\{P_\alpha\}_{\alpha \geq 1}$ as the leading terms, with respect to the order α of the Taylor expansions.

If we take other discrete dynamics $w_{N+1}^{t+1} = g(z_{N-l_0}^{t+1}, \ldots, z_{N+k_{d+1}}^{t-d})$, then one obtains two families of the partial differential equations $\{P_\alpha\}_{\alpha \geq 1}$ and $\{Q_\alpha\}_{\alpha \geq 1}$ correspondingly.

Definition 14.3. Two discrete dynamics:

$$z_{N+1}^{t+1} = f(z_{N-l_0}^{t+1}, \ldots, z_{N+k_{d+1}}^{t-d}),$$

$$w_{N+1}^{t+1} = g(z_{N-l_0}^{t+1}, \ldots, z_{N+k_{d+1}}^{t-d})$$

are infinitely unrelated in $\sim_*^+$, if for any constant M, c, D, L, there are some α so that P_α and Q_α are mutually unrelated in $\sim_*^+$ in the class $(M, c, D, L; \alpha)$.

Below we give pairs of PDEs which are mutually unrelated. We treat two cases:

(1) they are both in $\mathbf{PDE}_2^{fin}$, which are exponentially unrelated in $\sim_{fin}^e$. They arose from two discrete dynamics which are mutually infinitely unrelated.

(2) they are both in $\mathbf{PDE}_2^\infty$, which are double-exponentially unrelated in $\sim_{fin}^{e^e}$.

14.6 Unrelated pairs in the linear case

Below we show that the following linear equations are mutually unrelated:

Proposition 14.1. *For any $M, D, L > 1$, there are l_0 and a_0, b_0 so that for each even $l = 2m \geq l_0$, two PDEs $u, v : (0, a_0) \times [0, b_0) \to (0, \infty)$ of order l:*

$$\epsilon u_s + \frac{\epsilon^2}{2} u_{2s} + \cdots + \frac{\epsilon^l}{l!} u_{ls} + \epsilon^2 u_{xs} + \frac{\epsilon^3}{2} u_{xss} + \cdots + \frac{\epsilon^l}{l!} u_{(l-1)xs} = 0,$$

$$\epsilon(v_s + v_x) + \epsilon^2 \left(v_{xs} + \frac{1}{2} v_{2s} + \frac{1}{2} v_{2x}\right) + \cdots + \frac{\epsilon^l}{l!} v_{lx} = 0$$

satisfy the following;
(1) they are approximable in the class $(1, 1, 0, 1, 1)$ in $\mathbf{PDE}_2^{fin}$, and
(2) they are mutually unrelated in $(M, 1, D, L; l)$.

In particular they are mutually infinitely unrelated.

Proof. Step 1: Let us consider the discrete dynamics given by:

$$z_{N+1}^{t+1} = z_{N+1}^t,$$
$$w_{N+1}^{t+1} = w_N^t$$

and take the rescaling parameters by $N = \frac{x}{\epsilon}, t = \frac{s}{\epsilon}$ and $z_N^t = u(x, s), w_N^t = v(x, s)$. By taking the Taylor expansions up to order $l + 1$, one obtains the above pair of the partial differential equations in $\mathbf{PDE}_2^{fin}$, which are both approximable in $(1, 1, 0, 1, 1, 1)$. Notice that each monomial in the PDE in u contains derivatives by s.

Step 2: Let $f(x) = x^l + 1$, and put $u(x, s) = f(x)$ and $v(x, s) = f(x-s)$. It is immediate to see that they are solutions respectively, because of the independence of the variable s for u, and of the symmetry of the equation for v. Moreover the equalities $K(u)_{l+1} = K_{l+1}(v) = 0$ hold, since $l + 1$

derivatives of u and v are both equal to zero. We require that l is even in order to guarantee positivity of values of v.

Step 3: If they were related in $(M, 1, D, L; l)$, then there is some C independent of solutions so that they must satisfy the asymptotic estimates:

$$\left(\frac{u(x,s)}{v(x,s)}\right)^{\pm 1} \leq M^{\epsilon^{-D}(x+s+1)}[u:v]_{L\epsilon}$$

for all $0 < L\epsilon \leq \min(a_0, b_0, C)$.

Let us choose small $\epsilon > 0$ with $L\epsilon \leq 1$. Then the estimates hold:

$$M^{\epsilon^{-D}(x+s+1)} \leq M^{\epsilon^{-D}(a_0+b_0+1)},$$

$$[u:v]_{L\epsilon} \leq [u:v]_1 = \max\left(\sup_{0<x\leq a_0}\frac{x^l+1}{(x-1)^l+1}, \sup_{0<x\leq 1}\frac{(b_0-x)^l+1}{x^l+1}\right)$$

where:

$$\frac{x^l+1}{(x-1)^l+1} = \frac{1+\frac{1}{x^l}}{(1-\frac{1}{x})^l+\frac{1}{x^l}} \leq \begin{cases}\frac{2}{\frac{1}{2^l}} = 2^{l+1} & x \geq 2, \\ 2^{l+1} & x \leq 2.\end{cases},$$

$$\sup_{0<x\leq 1}\frac{(b_0-x)^l+1}{x^l+1} = b_0^l + 1.$$

Then the estimate $[u:v]_{L\epsilon} \leq b_0^l + 1$ holds, if we choose $b_0 \geq 3$.

Now let us find $a_0 > b_0 \geq 3$ so that the inequality $\frac{a_0}{a_0-b_0} > b_0$ holds. In fact for $2 < \beta < 4$ (say $\beta = 3$ is enough), if one chooses large a_0 with $a_0^2 - \beta a_0 \geq 0$ then

$$b_0 = \frac{1}{2}\left(a_0 + \sqrt{a_0^2 - \beta a_0}\right) \geq 3$$

satisfies the required conditions. Notice that the equality $\frac{a}{a-b} = b$ holds, in the case when $b = \frac{1}{2}(a + \sqrt{a^2 - 4a})$.

Let us fix such a pair (a_0, b_0). Now $\frac{u(a_0,b_0)}{v(a_0,b_0)} = \frac{a_0^l+1}{(a_0-b_0)^l+1}$ holds. If one chooses sufficiently large $l >> 1$, then the estimates:

$$\frac{u(a_0,b_0)}{v(a_0,b_0)}(b_0^l+1)^{-1} = \frac{a_0^l+1}{(a_0-b_0)^l+1}\frac{1}{(b_0^l+1)} \begin{cases}\geq \frac{1}{4}\frac{a_0^l}{b_0^l} & (a_0-b_0 \leq 1) \\ \geq \frac{1}{4}\frac{a_0^l}{(a_0-b_0)^l}\frac{1}{b_0^l} & (a_0-b_0 > 1)\end{cases}$$

$$> M^{\epsilon^{-D}(a_0+b_0+1)}$$

hold, since the right hand side is independent of l. Then one has:

$$\left(\frac{u(a_0,b_0)}{v(a_0,b_0)}\right)^{\pm 1} > M^{\epsilon^{-D}(a_0+b_0+1)}(b_0^l+1) \geq M^{\epsilon^{-D}(a_0+b_0+1)}[u:v]_{L\epsilon}.$$

This is a contradiction.

$\square$

14.7 Non linear estimates

Let us treat the case of double-exponential estimates. Let $P(\epsilon, u, u_s, u_x, \ldots, u_{\alpha s}) = 0$ be in $\mathbf{PDE}_2^\infty$ of order $\alpha \geq 1$, and let us compare its solutions with the translations:

$$P_I : v_s = Iv_x \qquad (I > 0).$$

Let us start from a general situation.

Lemma 14.3. *Suppose that for some $\frac{1}{4} \geq \delta_0 > 0$, $C, C', C'' > 0$ and for all small $0 < \epsilon \leq \epsilon_0$, there are solutions $u_\epsilon(x, s)$ on $(0, \delta_0) \times [0, \delta_0) \to (0, \infty)$ with the initial values $u(x, 0) = f(x) = 1 - x$, which satisfy both the bounds:*

(1) $C \leq u_\epsilon(x, s) \leq C'$, (2) $K_{\alpha+1}(u_\epsilon) \leq C''$.

Then for any $M, c, D, L \geq 1$, there is some $I_0 > 0$ so that for all $I \geq I_0$, P and P_I are unrelated in $\sim_{fin}^{e^e}$ in the class $(M, c, D, L; \alpha)$.

Proof. We verify the conclusion for a specific $I_0 > 0$. Then the general case follows by restriction of the domain of the solutions by replacing $\delta_0 > 0$ by a smaller one.

Let us put $v(x, s) = f(x + I_0 s)$. Then v is the solution to P_{I_0} and $K_{\alpha+1}(v) = 0$ holds, since f is linear.

Let us take a smaller $\delta_0 \gg \epsilon \gg \delta > 0$, and choose I_0 with $I_0 \delta_0 = 1 - \delta_0 - \delta$. One may assume the estimate $I_0 L\epsilon \leq \delta_0$. Then $\delta_0 + I_0 L\epsilon \leq 2\delta_0$ and $L\epsilon + I_0 \delta_0 = 1 - \delta_0 - \delta + L\epsilon \leq 1 - \frac{\delta_0}{2}$. So the estimate $x + I_0 s \leq 1 - \frac{\delta_0}{2}$ holds on the initial domain $(x, s) \in (0, \delta_0) \times [0, L\epsilon) \cup (0, L\epsilon) \times [0, \delta_0)$. Thus the estimate holds:

$$[v : u_\epsilon]_{L\epsilon} \leq 2C' \delta_0^{-1}.$$

Suppose they could be $(M, c, D, L; \alpha)$ related. Then they must satisfy the asymptotic estimates $(\frac{u_\epsilon(x,s)}{v(x,s)})^{\pm 1} \leq Mc^{2\epsilon^{-D}}(2C'\delta_0)^{c^{2\epsilon^{-D}}}$.

On the other hand $v(\delta_0, \delta_0) = \delta$ and $u_\epsilon(\delta_0, \delta_0) \geq C$ holds. So $\frac{u_\epsilon(\delta_0,\delta_0)}{v(\delta_0,\delta_0)} \geq \frac{C}{\delta}$ which can be arbitrarily large. This is a contradiction. $\square$

Let us apply the above situation to the quasi-linear equations and the translations:

$$4u_s + \epsilon u(u_s + u_x) = 0 , \qquad v_s = Iv_x.$$

We verify that both are in $\mathbf{PDE}_2^\infty$ and are mutually unrelated in $\sim_{fin}^{e^e}$ (Compare this with lemma 14.1):

Proposition 14.2. *For any M, c, D, L, there is I_0 so that for all $I \geq I_0$, the pair of the partial differential equations given by:*

$$4u_s + \epsilon u(u_s + u_x) = 0,$$
$$v_s = I v_x$$

is in $\mathbf{PDE}_2^\infty$ in $(20, 2, 1, I + 1, 1)$, and is unrelated in $\sim_{fin}^{e^c}$ in the class $(M, c, D, L; 1)$.

Before proceeding, let us briefly recall a way to produce solutions, called the *method of characteristics* for the conservative non linear equations of the form $u_s + F(u)u_x = 0$.

Let $u(\xi, 0) = f(\xi)$ be an initial condition, and try to solve the equation $\frac{dx}{ds} = F(u)$ for $u = u(x, s)$ with $x(\xi, 0) = \xi$. Then $\frac{du(x,s)}{ds} = u_s + \frac{dx}{ds} u_x = 0$ holds. So u is constant along $x(\xi, s)$. Moreover $\frac{d^2 x}{ds^2} = \frac{F(u)}{du} \frac{du}{ds} = 0$, and so $x(\xi, s) = F(f(\xi))s + \xi$. Thus if one could solve $\xi = \xi(s, x)$, then $f(\xi(x, s))$ will give us solutions, since u is constant along $x(\xi, s)$,

Proof. The second equation is induced from the discrete dynamics:

$$w_{N+1}^{t+1} = w_{N+I+1}^t$$

which lies in the class $(1, 1, 0, I + 1, 1)$. Thus both are in $\mathbf{PDE}_2^\infty$ in the class $(20, 2, 1, I + 1, 1)$ by lemma 14.2.

Now let us consider the initial function $f : (0, 1) \to (0, \infty)$ by $f(x) = 1 - x$. The conclusion follows if one can find solutions u which satisfy two conditions (1) and (2) in Lemma 14.3. We will solve the equation concretely using the method of characteristics.

Let us consider the equation:

$$x - \xi = \frac{\mu f(\xi)s}{1 + \mu f(\xi)} = \frac{\mu(1 - \xi)s}{1 + \mu(1 - \xi)}, \qquad (\, \mu = \frac{\epsilon}{4} \,).$$

Then it gives the equation $\mu \xi^2 - (1 + \mu(x - s + 1))\xi + (1 + \mu)x - \mu s = 0$, and one can solve it with $x(\xi, 0) = \xi$:

$$\xi(x, s) = \frac{1}{2\mu} \left[1 + \mu(x - s + 1) - \sqrt{(1 + \mu(x - s + 1))^2 - 4\mu\{(1 + \mu)x - \mu s\}} \right].$$

Let the solution $u(x, s) = 1 - \xi(x, s)$. Then for a small $\delta_0 > 0$ and all sufficiently small $\epsilon > 0$, two conditions (1) $C \leq u(x, s) \leq C'$ and (2) $K_2(u) \leq C''$ are certainly satisfied, by elementary calculations.

$\square$

Remark 14.5. Positivity of solutions follows from the structure of the equation. Let us rewrite the equation as $u_s = -\epsilon u u_x (4 + \epsilon u)^{-1}$. At $s = 0$, $u_x(x, 0) = -1 < 0$ holds and so $u_s(x, 0) = \epsilon u (4 + \epsilon u)^{-1} > 0$ holds. Thus there is some $T_0 > 0$ so that $u_s(x, s) > 0$ still holds for all $(x, s) \in (0, \frac{1}{2}) \times [0, T_0)$. In particular $u(x, s) \geq \frac{1}{2}$ for all $(x, s) \in (0, \frac{1}{2}) \times [0, T_0)$. The same argument works for the equation $v_s + \frac{\epsilon}{2} v v_x - \frac{1}{2} v^2 = 0$.

Now in particular one has obtained the following:

Corollary 14.2. *There are unrelated pairs both in*

$$(1) \ (\mathbf{PDE}_2^{fin}, \ \sim_{fin}^{e}), \quad (2) \ (\mathbf{PDE}_2^{\infty}, \ \sim_{fin}^{e^e})$$

with respect to any constant.

Finally we would like to address some problems:

Question 14.1. *Are there unrelated pairs in the following classes:*

$$(1) \ (\mathbf{PDE}_2^{fin}, \ \sim_{fin}^{e^e}), \quad (2) \ (\mathbf{PDE}_2^{\infty}, \ \sim_{\infty}^{e}), \quad (3) \ (\mathbf{PDE}_2^{\infty}, \ \sim_{\infty}^{e^e}).$$

14.8 Computational aspects

Our method of comparison is intimately suitable for the computer calculations, since all constants of the estimates are quite explicit. By using numerical calculations, we reprove proposition 14.1 for a particular case of the constant values with $M = 10^3$ and $l = 10^3$.

To see that two partial differential equations are unrelated, it is enough to find a test point which does not satisfy the inequality of the asymptotic estimates.

Let us try and compare numerical calculations in two cases of the related and the unrelated pairs. Recall that we have induced partial differential equations from discrete dynamics, and we compute these discrete dynamics.

Let $\{z_N^t\}_{N,t \geq 0}$ and $\{w_N^t\}_{N,t \geq 0}$ be two discrete dynamics. Then for L, N_0 and t_0, we let the discrete version of the initial rates be:

$$[\{z_N^t\} : \{w_N^t\}]_{L, N_0, t_0} \equiv \sup_{0 \leq N \leq N_0, 0 \leq t \leq t_0, 0 \leq a \leq L} \max \left\{ \left(\frac{z_N^a}{w_N^a}\right)^{\pm}, \left(\frac{z_a^t}{w_a^t}\right)^{\pm} \right\}.$$

Let us denote $\tilde{u}(\epsilon N, \epsilon t) = z_N^t$ and $\tilde{v}(\epsilon N, \epsilon t) = w_N^t$ respectively. Then we regard that both $\tilde{u}$ and $\tilde{v}$ approximate u and v respectively:

$$\tilde{u}(\epsilon N, \epsilon t) \sim u(\epsilon N, \epsilon t)$$

and similarly for v.

Below we choose the rescaling parameters $(m, p, q) = (1, 1, 1)$. All of the numerical calculations here are performed using the computer algebra system "Maple 13" with rational or floating number manipulations. Then the numbers after calculation are converted to the floating-point numbers for presentation purposes.

14.9 Case of the unrelated pair

Let us recall two partial differential equations in proposition 14.1:

$$\epsilon u_s + \frac{\epsilon^2}{2} u_{2s} + \cdots + \frac{\epsilon^l}{l!} u_{ls} + \epsilon^2 u_{xs} + \frac{\epsilon^3}{2} u_{xss} + \cdots + \frac{\epsilon^l}{l!} u_{(l-1)xs} = 0, \qquad (*_1)$$

$$\epsilon(v_s + v_x) + \epsilon^2(v_{xs} + \frac{1}{2} v_{2s} + \frac{1}{2} v_{2x}) + \cdots + \frac{\epsilon^l}{l!} v_{lx} = 0 \qquad (*_2)$$

which are mutually induced by the discrete dynamics:

$$z_{N+1}^{t+1} = z_{N+1}^t \ , \ w_{N+1}^{t+1} = w_N^t.$$

We have verified that the pair of the solutions $u(x, s) = x^l + 1$ and $v(x, s) = (x - s)^l + 1$ breaks the exponential inequality:

$$\max\left\{ \ \frac{u(x, s)}{v(x, s)} \ , \ \frac{v(x, s)}{u(x, s)} \ \right\} >> M^{\epsilon^{-D}(x+s+1)}[u : v]_{L\epsilon}.$$

Now for $N, t = 0, 1, 2, \ldots$, let us put the solutions to the above discrete dynamics:

$$z_N^t = (\epsilon N)^l + 1 \ , \ w_N^t = \epsilon^l(N - t)^l + 1$$

which are exactly the same as the solutions $u(x, s)$ and $v(x, s)$ by rescaling of the variables $x = N\epsilon$ and $s = t\epsilon$ respectively.

We let:

$$\tilde{Q}_e(x, s) = \log \max\{\frac{w_N^t}{z_N^t}, \frac{z_N^t}{w_N^t}\} - \log \ [\{z_N^t\} : \{w_N^t\}]_{L,\epsilon^{-1}x, \epsilon^{-1}s}$$

which takes much bigger values than $\epsilon^{-D}(x + s + 1) \log M$ as we have seen before. Let:

$$Q_e(x, s) = \max\left(0, \tilde{Q}_e(x, s)\right).$$

We calculate the values $Q_e(x, s)$ at the points $(x, s) = (N\epsilon, t\epsilon)$ for $\{N, t \in \mathbb{Z} \mid 1 < N \le A_0/\epsilon, 1 < t \le T_0/\epsilon\}$.

Now we verify the following by computer calculation. These calculations are performed by using rational numbers which means that the calculated values are exact ones.

Proposition 14.3. *The pair of the equations* $(*_1, *_2)$ *are mutually unrelated in* $\sim^e_{fin}$ *in* $(10^3, 1, 1, 1; 10^3)$ *at* $\epsilon = 1/2$.

Proof. Since the values of discrete dynamics z^t_N and w^t_N are precisely the same as the solutions $z^t_N = u(N\epsilon, t\epsilon)$ and $w^t_N = v(x, s) = (N\epsilon, t\epsilon)$ respectively, it is enough to verify that $Q_e(x, s)$ certainly hit bigger values than $\epsilon^{-D}(x + s + 1)\log M = 6(x + s + 1)\log 10$ at some points.

Let us choose constants by $l = 1000, L = 0$ and $\epsilon = 1/2$, and consider the values of $\tilde{Q}_e$ at each point $(x, s) = (N\epsilon, t\epsilon)$ for $1 \leq N, t \leq 8$. For example $\tilde{Q}_e(1.5, 2) = \log\max\{\frac{w^4_3}{z^4_3}, \frac{z^4_3}{w^4_3}\} - \log\left[\{z^t_N\} : \{w^t_N\}\right]_{0,3,4}$.

Then we show the values of $Q_e(x, s)$ with 4 digits of precision at the point $(x, s) = (N\epsilon, t\epsilon)$ for $\{N, t \in \mathbb{Z} \mid 1 < N, t \leq 8 = 4/\epsilon\}$ in Table 14.1.

Table 14.1   Computed values of $Q(N\epsilon, t\epsilon)$ (Unrelated pair)

$N\backslash t$	0	1	2	3	4	5	6	7	8
0	0.0	0.0	0.0	0.0	0.0	0.0	0.0	0.0	0.0
1	0.0	0.0	0.0	0.0	0.0	0.0	0.0	0.0	0.0
2	0.0	0.6931	0.0	0.0	0.0	0.0	0.0	0.0	0.0
3	0.0	404.8	404.8	0.0	0.0	0.0	0.0	0.0	0.0
4	0.0	287.7	691.8	287.6	0.0	0.0	0.0	0.0	0.0
5	0.0	223.1	510.1	510.1	223.2	0.0	0.0	0.0	0.0
6	0.0	182.3	404.8	287.6	404.9	182.7	0.0	0.0	0.0
7	0.0	154.2	335.8	154.1	154.2	335.7	154.0	0.0	0.0
8	0.0	135.5	287.0	64.50	0.0	64.50	287.0	133.0	0.0

Now for $D = 1, M = 1000$, $\log M^{\epsilon^{-D}(x+s+1)} \leq \log M^{\epsilon^{-D}(4+4+1)} \sim$ 124.3. So the table presents several points of (N, t) which break the exponential asymptotic estimates:

$$Q_e(x, s) >> \log M^{\epsilon^{-D}(x+s+1)}.$$

$\square$

14.10   Case of the related pair

In the case of the example of the related pair presented earlier, the numerical calculations do not directly provide the mathematical proof of the relevancy.

Nevertheless such calculations give us several insights on the actual behavior when compared with the case of the unrelated pair.

Let us recall two partial differential equations in lemma 14.2:

$$v_s + \frac{\epsilon}{2}vv_x - \frac{1}{2}v^2 = 0,$$

$$2u_s + \frac{\epsilon}{2}u(u_s + u_x) = 0$$

which are mutually induced from the discrete dynamics:

$$z_{N+1}^{t+1} = \frac{z_{N+2}^t}{2} + \frac{z_N^t(1 + 2z_{N-1}^{t+1})}{2(1 + z_N^t)}, \quad (**_1)$$

$$w_{N+1}^{t+1} = \frac{w_{N+2}^t}{2} + \frac{w_N^t(1 + w_{N-1}^{t+1})}{2(1 + w_N^t)} \quad (**_2).$$

We have induced the asymptotic estimate for a pair of positive solutions which admit some estimates on higher derivatives in theorem 13.2. It implies the estimate:

$$\log\left(\frac{u(x, s)}{v(x, s)}\right)^{\pm 1} - 2^{\epsilon^{-1}(x+2s)+3}\log([u:v]_{2\epsilon}) \le 2^{\epsilon^{-1}(x+2s)+4}\log 40.$$

Now let us consider the solutions to the above discrete dynamics $(**)$ with the initial and boundary values:

$$z_N^t = (\epsilon N)^l + 1 \, , \ w_N^t = \epsilon^l(N - t)^l + 1$$

respectively, where $(N, t) = \{0\} \times \mathbb{N} \cup \mathbb{N} \times \{0\}$.

Let $\tilde{u}$ and $\tilde{v}$ be as in 14.8, and $u, v : [0, A_0] \times [0, T_0]$ be the solutions with the initial values. Then as before we regard that both $\tilde{u}$ and $\tilde{v}$ approximate u and v respectively at the points $(x, s) = (N\epsilon^p, t\epsilon^q) = (N\epsilon, t\epsilon)$ for $\{N, t \in \mathbb{Z} \,|\, 1 < N \le A_0/\epsilon, 1 < t \le T_0/\epsilon\,\}$.

We let:

$$\tilde{Q}_{e^\epsilon}(x, s) = \log\max\{\frac{w_N^t}{z_N^t}, \frac{z_N^t}{w_N^t}\} - 2^{\epsilon^{-1}(x+2s)+3}\log\,[\{z_N^t\} : \{w_N^t\}]_{L, \epsilon^{-1}x, \epsilon^{-1}s}.$$

Here we calculate the values using floating numbers with 100 digits of precision $Q_{e^\epsilon}(x, s) = \max\left(0, \tilde{Q}_{e^\epsilon}(x, s)\right)$, where we choose constants $l = 1000, L = 1, \epsilon = 1/10$ and $A_0 = T_0 = 1$. In particular the domain is $\{(x, s) = (N\epsilon, t\epsilon) : 0 \le N, t \le 10\}$.

It turns out from the numerical calculations that all the entries are equal to 0. In particular the estimates:

$$Q_{e^\epsilon}(x, s) \le 2^{4+(x+2s)/\epsilon}\log 40$$

Table 14.2   Computed values of $Q_e(N\epsilon, t\epsilon)$ (Related pair)

$N\backslash t$	0	1	2	3	4	5	6	7	8	9	10
0	0.0	0.0	0.0	0.0	0.0	0.0	0.0	0.0	0.0	0.0	0.0
1	0.0	0.0	0.0	0.0	0.0	0.0	0.0	0.0	0.0	0.0	0.0
2	0.0	0.0	0.095	0.189	0.258	0.331	0.393	0.463	0.524	0.0	0.0
3	0.0	0.0	0.196	0.327	0.461	0.567	0.682	0.776	0.0	0.0	0.0
4	0.0	0.0	0.229	0.450	0.618	0.794	0.926	0.0	0.0	0.0	0.0
5	0.0	0.0	0.318	0.565	0.813	0.991	0.0	0.0	0.0	0.0	0.0
6	0.0	0.0	0.334	0.670	0.899	0.0	0.0	0.0	0.0	0.0	0.0
7	0.0	0.0	0.399	0.677	0.0	0.0	0.0	0.0	0.0	0.0	0.0
8	0.0	0.0	0.294	0.0	0.0	0.0	0.0	0.0	0.0	0.0	0.0
9	0.0	0.0	0.0	0.0	0.0	0.0	0.0	0.0	0.0	0.0	0.0
10	0.0	0.0	0.0	0.0	0.0	0.0	0.0	0.0	0.0	0.0	0.0

follow, which supports the asymptotic estimate for particular constants. Notice that the approximate value is given by:

$$2^{4+(x+2s)/\epsilon}\log 40|_{\epsilon=1/10, x=s=10\epsilon} \sim 6.337 \times 10^{10}.$$

So far we have checked that certainly double exponential estimates hold for these pairs. Next let us examine whether they might still satisfy the exponential estimates.

Let us consider the values of $Q_e(x,s) = \max\left(0, \tilde{Q}_{e^\circ}(x,s)\right)$, where:

$$\tilde{Q}_e(x,s) = \log\max\left\{\frac{w_N^t}{z_N^t}, \frac{z_N^t}{w_N^t}\right\} - \log\left[\{z_N^t\} : \{w_N^t\}\right]_{1, \epsilon^{-1}x, \epsilon^{-1}s}$$

with the same constants, $l = 1000, L = 1, \epsilon = 1/10$ and $A_0 = T_0 = 1$. Table 14.2 gives the result of numerical calculations.

Let us compare their values with:

$$10\log 40 = \epsilon^{-1}\log 40|_{\epsilon=1/10} \le \epsilon^{-1}(x+s+1)\log 40$$
$$\le \epsilon^{-1}(x+s+1)\log 40|_{\epsilon=1/10, x=s=10\epsilon} = 30\log 40$$

where the left and right hand sides are approximately 36.89 and 110.7 respectively. Thus the inequality above holds for any point $(N,t) \in [0,10] \times [0,10]$.

Remark 14.6. It might happen that in both related and unrelated cases, the numerical computations can present such situations rather clearly. It seems to work quite effectively to examine the relation between unknown pair of partial differential equations by numerical calculations.

References

There is a classification project of partial differential equations in [BGH] from the viewpoint of integrable systems. It consists of a systematic theory to treat PDEs in some invariant way. Our approach is rather different from that since it focuses on local theory. This chapter is from [KT].

PART 3
Mealy type dynamics

Chapter 15

Hyperbolic system of partial differential equations

Let us recall the state system of the rational dynamics given by elementary rational functions:

$$\begin{cases} z_i^{j+1} = f(w_i^j, z_i^j, \ldots, z_{i+\alpha}^j), \\ w_{i+1}^j = g(w_i^j, z_i^j, \ldots, z_{i+\beta}^j) \end{cases}$$

where the initial values are given by $z_i^0 = z_i, w_0^j = w^j > 0$.

We shall follow the same process as previous chapters to induce partial differential equations which are systems in this case. In general it depends on the equations whether global solutions exist or they are unique. In the case of the Mealy automata we have considered, they are in the class of first order hyperbolic systems, which we call *hyperbolic Mealy systems*.

We study the basic analysis of hyperbolic Mealy systems including the existence and uniqueness.

15.1 Summary

It would be of interest to study how the characteristic properties of automata group actions on trees can be transmitted to the global behavior of solutions to the associated systems of partial differential equations.

Let $f_t = \frac{a_t}{b_t}$ and $g_t = \frac{c_t}{d_t}$ be two elementary rational functions, and consider the state systems of the rational dynamics:

$$\begin{cases} z_i^{j+1} = f_t(w_i^j, z_i^j, \ldots, z_{i+\alpha}), \\ w_{i+1}^j = g_t(w_i^j, z_i^j, \ldots, z_{i+\beta}^j). \end{cases}$$

Let the change of variables be:

$$i = \frac{x}{\epsilon}, \quad j = \frac{s}{\epsilon}, \quad u(x,s) = z_i^j \quad v(x,s) = w_i^j.$$

Then we take the difference, and insert the Taylor expansions:

$$z_i^{j+1} - f_t(w_i^j, z_i^j, \ldots, z_{i+\alpha}^j)$$
$$= \frac{P_1(\epsilon, t, u, v, u_s, u_x, \ldots, u_{\mu x}) + R_1(\epsilon, t, u, v, \ldots, u_{(\mu+1)x}(\xi))}{b_t(v(x, s), u(x, s), \ldots, u(x + \alpha\epsilon, s))}$$

where P_1 and R_1 are polynomials, and each monomial in R_1 contains derivatives of u of order $\mu + 1$.

Similarly we have the expansions:

$$w_{i+1}^j - g_t(w_i^j, z_i^j, \ldots, z_{i+\beta}^j) = v(x + \epsilon, s) - g_t(v(x, s), u(x, s), \ldots, u(x + \beta\epsilon, s))$$
$$= \frac{P_2(\epsilon, t, u, v, u_x, v_x, \ldots, u_{\mu x}) + R_2(\epsilon, t, u, v, \ldots, u_{(\mu+1)x}(\xi'))}{d_t(v(x, s), u(x, s), \ldots, u(x + \beta\epsilon, s))}.$$

Let us call the parametrized systems of PDE of order μ:

$$\begin{cases} P_1(\epsilon, t, u, v, u_s, \ldots, u_{\mu s}, u_{\mu x}) = 0, \\ P_2(\epsilon, t, u, v, v_x, \ldots, u_{\mu x}, v_{\mu x}) = 0 \end{cases}$$

the *induced systems of partial differential equations*.

Let us take four elementary rational functions f^l and g^l for $l = 1, 2$ so that $f^1 \sim f^2$ and $g^1 \sim g^2$ are tropically equivalent mutually. Let P_1^l and P_2^l be the defining polynomials for the induced systems of partial differential equations of order μ.

For functions of $C^{\mu+1}$ class, we will later introduce the *higher distortion* $K(u, v)$ which are the relatively uniform norms of their $\mu + 1$-differentials.

Let us state our asymptotic comparison estimate. Assume that the error terms are admissible.

Theorem 15.1. *For $l = 1, 2$, let C be the error constants.*

Let $(u^l, v^l) : [0, \infty) \times [0, \infty) \to (0, \infty)$ be positive solutions to the above systems respectively, so that the estimates:

$$0 \leq CK(u^l, v^l) \leq (1 - \delta)\epsilon^{-1-\mu}$$

are satisfied for some positive $\delta > 0$.

Then they satisfy the asymptotic estimates for all $(x, s) \in [0, \infty) \times [0, \infty)$:

$$\left(\frac{u^1}{u^2}\right)^{\pm 1}(x, s), \quad \left(\frac{v^1}{v^2}\right)^{\pm 1}(x, s)$$

$$\leq (N_0 M)^{10 P_{\epsilon^{-1}(x+s(\gamma+1))}(c)} \left([(u^1, v^1) : (u^2, v^2)]_\epsilon\right)^{3\tilde{c}^{\epsilon^{-1}(x+s(\gamma+1)+1)}}$$

where N_0 is any integer with $N_0 \geq \max(\delta^{-1}, 2 - \delta)$.

Among the induced systems of partial differential equations, the hyperbolic Mealy systems which we treat below are the ones which arise from automata semi-groups.

For Mealy automata given by:

$$\begin{cases} x(i, j+1) = \psi(y(i,j), x(i,j)), \\ y(i+1, j) = \phi(y(i,j), x(i,j)) \end{cases}$$

the induced first order systems of the equations:

$$\begin{cases} \epsilon\, u_s = f_t(v, u) - u, \\ \epsilon\, v_s = g_t(v, u) - v \end{cases}$$

are called the *hyperbolic Mealy systems*.

In practice, if we try to apply the above result, there appear two conditions which are:

(1) existence of positive solutions,

(2) estimates on higher distortions.

Below we develop a basic PDE analysis of the hyperbolic Mealy systems. In particular we verify the existence of positive solutions, uniqueness and explicit C^1 estimates. Construction of admissible solutions to the hyperbolic Mealy systems involve the interplay of estimates between piecewise linear and differentiable dynamics.

Theorem 15.2. *Suppose that the pair (f_t, g_t) restricts to a self-dynamics over $[r, R]$, and give the positive initial values:*

$$u, \ v : \ I_0 \to [r + q, R - q]$$

where $I_0 = [0, \infty) \times \{0\} \cup \{0\} \times [0, \infty)$ is the initial domain.
Then: (1) there exists a positive solution

$$u, v : [0, \infty) \times [0, \infty) \to (0, \infty)$$

with the uniform bounds $r + q \le u(x, s)$, $v(x, s) \le R - q$.
(2) The solution is unique.

We also induce the C^1 estimates which are well known for the hyperbolic systems, but here we also estimate the constants explicitly.

We construct admissible systems of partial differential equations from piecewisely linear functions. For this purpose we will later introduce some modification of automata, called *refinement*. Let **A** be a Mealy automaton with 2 alphabets, equipped with a representative (ψ, ϕ) by relative (max, +)-functions.

Theorem 15.3. *For any $C > 0$ and any $t \geq t(C) > 1$, there are refinements $(\bar{\psi}, \bar{\phi})$ of (ψ, ϕ) with the pairs of tropical correspondences $(\bar{f}_t, \bar{g}_t)$ so that $(\bar{f}_t, \bar{g}_t)$ admits admissible solutions.*

15.2   Rough approximation by discrete dynamics

Let us describe our general procedure to approximate solutions to the systems of partial differential equations by state systems of rational dynamics.

Let $f_t = \frac{a_t}{b_t}$ and $g_t = \frac{c_t}{d_t}$ be elementary rational functions of $\alpha + 2$ and $\beta + 2$ variables respectively, where a_t, b_t, c_t, d_t are elementary polynomials. Let us consider the state systems of the rational dynamics:

$$\begin{cases} z_i^{j+1} = f_t(w_i^j, z_i^j, \ldots, z_{i+\alpha}^j), \\ w_{i+1}^j = g_t(w_i^j, z_i^j, \ldots, z_{i+\beta}^j). \end{cases}$$

For a constant $0 < \epsilon \leq 1$, let us consider a $C^{\mu+1}$ function $u : [0, \infty) \times [0, \infty) \to (0, \infty)$, and take the Taylor expansions:

$$u(x + i\epsilon, s + j\epsilon) = u + i\epsilon u_x + j\epsilon u_s + \frac{(i\epsilon)^2}{2} u_{2x} + \frac{(j\epsilon)^2}{2} u_{2s}$$

$$+ ij\epsilon^2 u_{xs} + \cdots + \frac{(i\epsilon)^\mu}{\mu!} u_{\mu x} + \frac{(j\epsilon)^\mu}{\mu!} u_{\mu s}$$

$$+ \frac{(i\epsilon)^{(\mu+1)}}{(\mu+1)!} u_{(\mu+1)x}(\xi_{\alpha+1,0}) + \cdots + \frac{(j\epsilon)^{(\mu+1)}}{(\mu+1)!} u_{(\mu+1)s}(\xi_{0,\alpha+1})$$

where $|(x, s) - \xi_{ij}| \leq (|i| + |j|)\epsilon$ holds.

Let us introduce the change of variables by:

$$i = \frac{x}{\epsilon}, \quad j = \frac{s}{\epsilon}, \quad u(x, s) = z(i, j), \quad v(x, s) = w(i, j).$$

Then we take the difference, and insert the Taylor expansions:

$$z_i^{j+1} - f_t(w_i^j, z_i^j, \ldots, z_{i+\alpha}^j)$$

$$= u(x, s + \epsilon) - f_t(v(x, s), u(x, s), \ldots, u(x + \alpha\epsilon, s))$$

$$= \frac{P_1(\epsilon, t, u, v, u_x, v_x, \ldots, v_{\mu s}) + R_1(\epsilon, t, u, v, \ldots, v_{(\mu+1)s}(\xi'_{0,\mu+1}))}{b_t(v(x, s), u(x, s), \ldots, u(x + \alpha\epsilon, s))}$$

$$\equiv \mathbf{L}_1(\epsilon, t, u, v, u_x, \ldots, v_{\mu s})$$

$$+ \epsilon^{\mu+1} \mathbf{E}_1(\epsilon, t, u, v, \ldots, \{u_{\bar{a}}(\xi_{\bar{a}})\}_{\bar{a}}, \{v_{\bar{b}}(\xi'_{\bar{b}})\}_{\bar{b}})$$

where P_1 and R_1 are polynomials, and each monomial in R_1 contains derivatives of u of order $\mu + 1$.

Similarly we have the expansions:

$$w_{i+1}^j - g_t(w_i^j, z_i^j, \ldots, z_{i+\beta}^j)$$

$$= v(x + \epsilon, s) - g_t(v(x, s), u(x, s), \ldots, u(x + \beta\epsilon, s))$$

$$= \frac{P_2(\epsilon, t, u, v, u_x, v_x, \ldots, v_{\mu s}) + R_2(\epsilon, t, u, v, \ldots, v_{(\mu+1)s}(\xi'_{0,\mu+1}))}{d_t(v(x, s), u(x, s), \ldots, u(x + \beta\epsilon, s))}$$

$$\equiv \mathbf{L}_2(\epsilon, t, u, v, u_x, v_x, \ldots, v_{\mu s})$$

$$+ \epsilon^{\mu+1} \mathbf{E}_2(\epsilon, t, u, v, \ldots, \{u_{\bar{a}}(\xi_{\bar{a}})\}_{\bar{a}}, \{v_{\bar{b}}(\xi'_{\bar{b}})\}_{\bar{b}})$$

where each monomial in R_1 contains derivatives of u or v of order $\mu + 1$.

We say that $\mathbf{L}_i$ and $\mathbf{E}_i$ are the *leading* and *error* terms respectively.

Once one has chosen a pair of elementary rational functions (f_t, g_t), then the above process determines a system of the partial differential equations:

$$\begin{cases} P_1(\epsilon, t, u, v, u_x, u_s, u_{2x}, \ldots, u_{\mu x}) = 0 \\ P_2(\epsilon, t, v, u, u_x, v_x, u_{2x}, \ldots, u_{\mu x}) = 0 \end{cases}$$

which are called the *induced systems of partial differential equations* of order μ.

Tropical geometry provides an automaton given by $(\max, +)$-functions (ψ, ϕ). So the pair (f_t, g_t) plays the role of a bridge to connect the automata and the systems of partial differential equations.

15.2.1 *Mealy automaton*

Let (ψ, ϕ) be a pair of relative $(\max, +)$-functions of two variables, and consider the Mealy dynamics given by:

$$\begin{cases} x_i^{j+1} = \psi(y_i^j, x_i^j), \\ y_{i+1}^j = \phi(y_i^j, x_i^j). \end{cases}$$

Let (f, g) be elementary rational functions corresponding to (ψ, ϕ) respectively, and consider the systems of the rational dynamics:

$$\begin{cases} z_i^{j+1} = f(w_i^j, z_i^j), \\ w_{i+1}^j = g(w_i^j, z_i^j). \end{cases}$$

Definition 15.1. The induced first order system of the equations:

$$\begin{cases} \epsilon\, u_s = f(v, u) - u, \\ \epsilon\, v_s = g(v, u) - v \end{cases}$$

is called the hyperbolic Mealy system.

Notice that the error terms are given by the following, respectively:

$$\mathbf{E}_1 = \frac{1}{2}\frac{\partial^2 u}{\partial s^2}, \quad \mathbf{E}_2 = \frac{1}{2}\frac{\partial^2 v}{\partial x^2}$$

which are always admissible.

15.3 Higher distortion and asymptotic comparison

Let f_t and g_t be elementary rational functions of $\alpha + 2$ and $\beta + 2$ variables respectively.

Let $u, u', v, v' : [0, \infty) \times [0, \infty) \to (0, \infty)$ be four functions. The *initial rate of the pair* with respect to $\epsilon > 0$ is defined by:

$$[(u, v) : (u', v')]_\epsilon$$

$$= \max \Big[\sup_{(x,s)\in[0,\infty)\times[0,\epsilon]} \max\Big\{\frac{u}{u'}(x,s), \frac{u'}{u}(x,s)\Big\} ,$$

$$\sup_{(x,s)\in[0,\epsilon]\times[0,\infty)} \max\Big\{\frac{v}{v'}(x,s), \frac{v'}{v}(x,s)\Big\} \Big].$$

Let $(u, v) : [0, \infty) \times [0, \infty) \to (0, \infty)^2$ be a pair of functions of $C^{\mu+1}$ class. We introduce the pointwise norms by:

$$||(u,v)||^1_{\mu,\alpha}(x,s) = \begin{cases} \max\Big[\sup_{(x,s)\in[x,x+\alpha\epsilon]\times\{s\}} \big|\frac{\partial^{\mu+1}u}{\partial^{\mu+1}_x}\big|(x,s), \\ \quad \sup_{(x,s)\in\{x\}\times[s,s+\epsilon]} \big|\frac{\partial^{\mu+1}u}{\partial^{\mu+1}_s}\big|(x,s)\Big], & \alpha \geq 1, \\ \\ \sup_{(x,s)\in\{x\}\times[s,s+\epsilon]} \big|\frac{\partial^{\mu+1}u}{\partial^{\mu+1}_s}\big|(x,s), & \alpha = 0, \end{cases}$$

$$||(u,v)||^2_{\mu,\beta}(x,s) = \begin{cases} \max\Big[\sup_{(x,s)\in[x,x+\beta\epsilon]\times\{s\}} \big|\frac{\partial^{\mu+1}u}{\partial^{\mu+1}_x}\big|(x,s), \\ \quad \sup_{(x,s)\in[x,x+\epsilon]\times\{s\}} \big|\frac{\partial^{\mu+1}v}{\partial^{\mu+1}_x}\big|(x,s)\Big], & \beta \geq 1, \\ \\ \sup_{(x,s)\in[x,x+\epsilon]\times\{s\}} \big|\frac{\partial^{\mu+1}v}{\partial^{\mu+1}_x}\big|(x,s), & \beta = 0. \end{cases}$$

Definition 15.2. Suppose the error terms are admissible. The higher distortion at order $\mu + 1$ is given by:

$$K(u, v) \equiv \sup_{(x,s)\in[0,\infty)^2} \max \Big[\frac{||(u,v)||^1_{\mu,\alpha}}{u(x, s + \epsilon)}, \frac{||(u,v)||^2_{\mu,\beta}}{v(x + \epsilon, s)} \Big].$$

The *error constant*:

$$C = \max(C_1, C_2)$$

is the number so that the pointwise estimates hold for any function u, v independently of ϵ and t:

$$
\begin{cases}
|\mathbf{E}_1(\epsilon, t, u, v, \dots, \{u_{\bar{a}}(\xi_{\bar{a}})\}_{\bar{a}})|(x, s) \leq C_1 ||(u, v)||^1_{\mu, \alpha}(x, s), \\
\\
|\mathbf{E}_2(\epsilon, t, u, v, \dots, \{u_{\bar{a}}(\xi_{\bar{a}})\}_{\bar{a}}, \{v_{\bar{a}}(\xi'_{\bar{a}})\}_{\bar{a}})|(x, s) \leq C_2 ||(u, v)||^2_{\mu, \beta}(x, s).
\end{cases}
$$

Example 15.1. For the Mealy case, the error constant is $C = \frac{1}{2}$, and we have the following:

$$
\begin{cases}
||(u, v)||^1_{\mu, \alpha} = \sup_{(x,s) \in \{x\} \times [s, s+\epsilon]} |\frac{\partial^2 u}{\partial^2 s}|(x, s), \\
||(u, v)||^2_{\mu, \beta} = \sup_{(x,s) \in [x, x+\epsilon] \times \{s\}} |\frac{\partial^2 v}{\partial^2 x}|(x, s).
\end{cases}
$$

Let us take four elementary rational functions f^1, f^2 and g^1, g^2 so that $f^1 \sim f^2$ and $g^1 \sim g^2$ are tropically equivalent mutually. For $l = 1, 2$, let:

$$
\begin{cases}
P^l_1(\epsilon, t, u, v, u_s, \dots, u_{\mu x}) = 0, \\
P^l_2(\epsilon, t, u, v, v_x, \dots, u_{\mu x}) = 0
\end{cases}
$$

be the induced systems of PDEs of order μ.

Assume that the error terms are admissible, and let M be the largest number of their components.

Theorem 15.4. *For $l = 1, 2$, let C be the error constants.*

Let $(u^l, v^l) : [0, \infty) \times [0, \infty) \to (0, \infty)$ be the solutions to the above systems respectively, so that the estimates:

$$0 \leq CK(u^l, v^l) \leq (1 - \delta)\epsilon^{-1-\mu}$$

are satisfied for some positive $\delta > 0$.

Then they satisfy the asymptotic estimates for all $(x, s) \in [0, \infty) \times [0, \infty)$:

$$\left(\frac{u^1}{u^2}\right)^{\pm 1}(x, s), \quad \left(\frac{v^1}{v^2}\right)^{\pm 1}(x, s)$$

$$\leq (N_0 M)^{10 P_{\epsilon^{-1}(x+s(\gamma+1))}(c)} \left([(u^1, v^1) : (u^2, v^2)]_\epsilon\right)^{3\bar{c}^{-1}(x+s(\gamma+1)+1)}.$$

where N_0 is any integer with $N_0 \geq \max(\delta^{-1}, 2 - \delta)$.

The key idea to the proof is also the dynamical inequalities for state systems of rational dynamics, which we verify below.

15.4 Dynamical inequalities

Let (ψ^1, ϕ^1) and (ψ^2, ϕ^2) be two pairs of relative $(\max, +)$-functions so that $\psi^1 \sim \psi^2$ and $\phi^1 \sim \phi^2$ are pairwisely equivalent. Let (f_t^1, g_t^1) and (f_t^2, g_t^2) be the pairs of the corresponding elementary rational functions.

Let us take the initial data $\{w^j\}_{j \geq 0}$ and $\{z_i\}_{i \geq 0}$ by positive numbers, and consider the solutions to the state systems of the rational dynamics:

$$\begin{cases} F_i^{j+1}(l) = f_t^l(G_i^j(l), F_i^j(l), \dots, F_{i+\alpha}^j(l)), \\ G_{i+1}^j(l) = g_t^l(G_i^j(l), F_i^j(l), \dots, F_{i+\beta}^j(l)), \end{cases}$$

with $F_i^0(l) = z_i$ and $G_0^j(l) = w^j$ for $l = 1, 2$.

Proposition 15.1. *Suppose two sequences $\{w_i^j\}_{i,j}$ and $\{z_i^j\}_{i,j}$ by positive numbers satisfy the dynamical inequalities:*

$$\begin{cases} f_t^1(w_i^j, z_i^j, \dots, z_{i+\alpha}^j) \leq z_i^{j+1} \leq f_t^2(w_i^j, z_i^j, \dots, z_{i+\alpha}^j) \\ g_t^1(w_i^j, z_i^j, \dots, z_{i+\beta}^j) \leq w_{i+1}^j \leq g_t^2(w_i^j, z_i^j, \dots, z_{i+\beta}^j) \end{cases}$$

with the same initial values $z_i^0 = z_i$ and $w_0^j = w^j$.

Then the uniform estimates hold for $l = 1, 2$:

$$\left(\frac{F_i^{j+1}(l)}{z_i^{j+1}}\right)^{\pm 1}, \quad \left(\frac{G_{i+1}^j(l)}{w_{i+1}^j}\right)^{\pm 1} \leq M^{2P_{i+j(\gamma+1)}(c)}.$$

Proof. Let $p_i^j = \log_t z_i^j$ and $q_i^j = \log_t w_i^j$. Then since $\log_t$ are monotone increasing, the estimates:

$$\begin{cases} \psi_t^1(q_i^j, p_i^j, \dots, p_{i+\alpha}^j) \leq p_i^{j+1} \leq \psi_t^2(q_i^j, p_i^j, \dots, p_{i+\alpha}^j) \\ \phi_t^1(q_i^j, p_i^j, \dots, p_{i+\beta}^j) \leq q_{i+1}^j \leq \phi_t^2(q_i^j, p_i^j, \dots, p_{i+\beta}^j). \end{cases}$$

hold by proposition 2.1, where $p_i^0 = x_i = \log_t z_i$ and $q_0^j = y^j = \log_t w^j$.

Let:

$$\begin{cases} x_l'(i, j) = \log_t F_i^j(l), \\ y_l'(i, j) = \log_t G_i^j(l). \end{cases}$$

Then $x_l'(i, 0) = p_i^0$ and $y_l'(0, j) = q_0^j$ hold, and they satisfy the equations for $l = 1, 2$:

$$\begin{cases} x_l'(i, j+1) = \psi_t^l(y_l'(i, j), x_l'(i, j), \dots, x_l'(i+\alpha, j)) \\ y_l'(i+1, j) = \phi_t^l(y_l'(i, j), x_l'(i, j), \dots, x_l'(i+\beta, j)). \end{cases}$$

We claim that the uniform estimates hold for $l = 1, 2$:

$$|x_l'(i, j+1) - p_i^{j+1}|, \ |y_l'(i+1, j) - q_{i+1}^j| \leq 2P_{i+j(\gamma+1)}(c) \log_t M.$$

Firstly suppose $p_i^{j+1} \geq x_1(i, j+1)$ holds. Then the estimates hold:

$$0 \leq |p_i^{j+1} - x_1'(i, j+1)| = p_i^{j+1} - x_1'(i, j+1)$$
$$\leq \psi_t^2(q_i^j, p_i^j, \ldots, p_{i+\alpha}^j) - \psi_t^1(y_1'(i, j), x_1'(i, j), \ldots, x_1'(i+\alpha, j))$$
$$\leq 2 \log_t M + |\psi^2(q_i^j, p_i^j, \ldots, p_{i+\alpha}^j) - \psi^1(y_1'(i, j), \ldots, x_1'(i+\alpha, j))|$$
$$\leq 2 \log_t M + c \max(|q_i^j - y_1'(i, j)|, |p_i^j - x_1'(i, j)|, \ldots, |p_{i+\alpha}^j - x_1'(i+\alpha, j)|)$$

by lemma 3.4. Conversely suppose $p_i^{j+1} \leq x_1'(i, j+1)$ holds. Then the estimates hold:

$$0 \leq |x_1'(i, j+1) - p_i^{j+1}| = x_1'(i, j+1) - p_i^{j+1}$$
$$\leq \psi_t^1(q_i^j, p_i^j, \ldots, p_{i+\alpha}^j) - \psi_t^1(y_1'(i, j), x_1'(i, j), \ldots, x_1'(i+\alpha, j))$$
$$\leq 2 \log_t M + |\psi^1(q_i^j, p_i^j, \ldots, p_{i+\alpha}^j) - \psi^1(y_1'(i, j), \ldots, x_1'(i+\alpha, j))|$$
$$\leq 2 \log_t M + c \max(|q_i^j - y_1'(i, j)|, |p_i^j - x_1'(i, j)|, \ldots, |p_{i+\alpha}^j - x_1'(i+\alpha, j)|).$$

So in all cases, the estimates hold:

$$|x_1'(i, j+1) - p_i^{j+1}| \leq 2 \log_t M + c \max(|q_i^j - y_1'(i, j)|, \ldots, |p_{i+\alpha}^j - x_1'(i+\alpha, j)|)$$

By replacing x_1 by x_2, one obtains the estimates:

$$|x_2'(i, j+1) - p_i^{j+1}| \leq 2 \log_t M + c \max(|q_i^j - y_2'(i, j)|, \ldots, |p_{i+\alpha}^j - x_2'(i+\alpha, j)|).$$

In the same way one obtains the estimates for $l = 1, 2$:

$$|y_l'(i+1, j) - q_{i+1}^j|$$
$$\leq 2 \log_t M + c \max(|q_i^j - y_l'(i, j)|, |p_i^j - x_l'(i, j)|, \ldots, |p_{i+\beta}^j - x_l'(i+\beta, j)|).$$

Applying proposition 6.1 for $x_i^j = x_l'(i, j)$ and $y_i^j = y_l'(i, j)$ with $T = \log_t M^2$, one obtains the estimates:

$$|x_l'(i, j+1) - p_i^{j+1}|, \ |y_l'(i+1, j) - q_{i+1}^j| \ \leq \ 2P_{i+j(\gamma+1)}(c) \log_t M.$$

Thus we have verified the claim.

Then one has the inequalities:

$$|\log_t \frac{z_i^{j+1}}{F_i^{j+1}(l)})|, \ |\log_t \frac{w_{i+1}^j}{G_{i+1}^j(l)})| \leq \log_t M^{2P_{i+j(\gamma+1)}(c)}.$$

By removing $\log_t$ from both sides, the conclusion follows.

$\square$

Let us consider the case when the initial data take different values. Recall $\tilde{c} = \max(1, c)$.

Let us consider the solutions to the state systems:

$$\begin{cases} F_i^{j+1}(l) = f_t^l(G_i^j(l), F_i^j(l), \dots, F_{i+\alpha}^j(l)), \\ G_{i+1}^j(l) = g_t^l(G_i^j(l), F_i^j(l), \dots, F_{i+\beta}^j(l)) \end{cases}$$

with the initial values $F_i^0(l) = z_i(1)$ and $G_0^j(l) = w^j(1)$.

Corollary 15.1. *Suppose other sequences $\{w_i^j\}_{i,j}$ and $\{z_i^j\}_{i,j}$ satisfy the dynamical inequalities:*

$$\begin{cases} f_t^1(w_i^j, z_i^j, \dots, z_{i+\alpha}^j) \leq z_i^{j+1} \leq f_t^2(w_i^j, z_i^j, \dots, z_{i+\alpha}^j), \\ g_t^1(w_i^j, z_i^j, \dots, z_{i+\beta}^j) \leq w_{i+1}^j \leq g_t^2(w_i^j, z_i^j, \dots, z_{i+\beta}^j) \end{cases}$$

with $z_i^0 = z_i(2)$ and $w_0^j = w^j(2)$.

Then the uniform estimates hold for $l = 1, 2$:

$$\left(\frac{F_i^{j+1}(l)}{z_i^{j+1}}\right)^{\pm 1}, \quad \left(\frac{G_{i+1}^j(l)}{w_{i+1}^j}\right)^{\pm 1}$$

$$\leq M^{4P_{i+j(\gamma+1)}(c)}[\{(z_i(k), w^j(k))\}_{k=1}^2]^{\tilde{c}^{i+1+j(\gamma+1)}}.$$

Proof. Let us consider another solution to the systems of the equations:

$$\begin{cases} J_i^{j+1}(l) = f_t^l(K_i^j(l), J_i^j(l), \dots, J_{i+\alpha}^j(l)), \\ K_{i+1}^j(l) = g_t^l(K_i^j(l), J_i^j(l), \dots, J_{i+\beta}^j(l)) \end{cases}$$

with the initial values $J_i^0(l) = z_i(2)$ and $K_0^j(l) = w^j(2)$. Then by proposition 6.3, the estimates hold for $l = 1, 2$:

$$\left(\frac{F_i^{j+1}(l)}{J_i^{j+1}(l)}\right)^{\pm 1}, \quad \left(\frac{G_{i+1}^j(l)}{K_{i+1}^j(l)}\right)^{\pm 1} \leq M^{2P_{i+j(\gamma+1)}(c)}[\{(z_i(k), w^j(k))\}_{k=1}^2]^{\tilde{c}^{i+1+j(\gamma+1)}}.$$

On the other hand by proposition 15.1, the estimates hold:

$$\left(\frac{J_i^{j+1}(l)}{z_i^{j+1}}\right)^{\pm 1}, \quad \left(\frac{K_{i+1}^j(l)}{w_{i+1}^j}\right)^{\pm 1} \leq M^{2P_{i+j(\gamma+1)}(c)}.$$

Then by multiplying both sides, one obtains the desired estimates:

$$\left(\frac{F_i^{j+1}(l)}{z_i^{j+1}}\right)^{\pm 1} = \left(\frac{J_i^{j+1}(l)}{z_i^{j+1}}\right)^{\pm 1} \left(\frac{F_i^{j+1}(l)}{J_i^{j+1}(l)}\right)^{\pm 1}$$

$$\leq M^{2P_{i+j(\gamma+1)}(c)} M^{2P_{i+j(\gamma+1)}(c)}[\{(z_i(k), w^j(k))\}_{k=1}^2]^{\tilde{c}^{i+1+j(\gamma+1)}}$$

$$= M^{4P_{i+j(\gamma+1)}(c)}[\{(z_i(k), w^j(k))\}_{k=1}^2]^{\tilde{c}^{i+1+j(\gamma+1)}}.$$

The desired estimates for $\left(\frac{G_{i+1}^j(l)}{w_{i+1}^j}\right)^{\pm 1}$ are obtained in the same way. $\square$

Proof of theorem 15.4.

For $0 \leq a, b \leq 1$, let the domain lattices be:

$$L_\epsilon(a, b) = \{(\,(l_1 + a)\epsilon, (l_2 + b)\epsilon\,) \in [0, \infty) \times [0, \infty) : l_1, l_2 \in \mathbb{N}\}.$$

Let:

$$\begin{cases} z_i^j(l) = u^l((i+a)\epsilon, (j+b)\epsilon), \\ w_i^j(l) = v^l((i+a)\epsilon, (j+b)\epsilon) \end{cases}$$

and consider the Taylor expansions:

$$\begin{cases} z_i^{j+1}(l) - f_t^l(w_i^j(l), z_i^j(l), \ldots, z_{i+\alpha}^j(l)) \\ \qquad = \mathbf{L}_1(\epsilon, t, u^l, v^l, u_s^l, \ldots, v_{\mu s}^l) \\ \qquad \qquad + \epsilon^{\mu+1} \mathbf{E}_1(\epsilon, t, u^l, v^l, \ldots, \{v_{\bar{a}}^l(\xi_{\bar{a}})\}_{\bar{a}}), \\ w_{i+1}^j(l) - g_t(w_i^j(l), z_i^j(l), \ldots, z_{i+\beta}^j(l)) \\ \qquad = \mathbf{L}_2(\epsilon, t, u^l, v^l, u_x^l, v_x^l, \ldots, v_{\mu s}^l) \\ \qquad \qquad + \epsilon^{\mu+1} \mathbf{E}_2(\epsilon, t, u^l, v^l, \ldots, \{u_{\bar{a}}^l(\xi_{\bar{a}}')\}_{\bar{a}}, \{v_{\bar{a}}^l(\xi_{\bar{a}}')\}_{\bar{a}}). \end{cases}$$

By assumption, the pairs (u^l, v^l) satisfy the systems of the equations for $l = 1, 2$:

$$\begin{cases} \mathbf{L}_1^l(\epsilon, t, u^l, v^l, u_x^l, \ldots) = 0, \\ \mathbf{L}_2^l(\epsilon, t, u^l, v^l, u_x^l, \ldots) = 0. \end{cases}$$

Moreover the error terms satisfy the estimates:

$$\begin{cases} |\mathbf{E}_1^l(\epsilon, t, u^l, v^l, \ldots)|(x, s) \leq CK(u^l, v^l)\, u^l(x, s + \epsilon), \\ |\mathbf{E}_2^l(\epsilon, t, u^l, v^l, \ldots)|(x, s) \leq CK(u^l, v^l)\, v^l(x + \epsilon, s). \end{cases}$$

Combining these, one obtains the estimates:

$$\begin{cases} |z_i^{j+1}(l) - f_t^l(w_i^j(l), z_i^j(l), \ldots, z_{i+\alpha}^j(l))| \\ \qquad \leq \epsilon^{\mu+1} |\mathbf{E}_1(\epsilon, t, u^l, v^l, \ldots,)| \\ \qquad \qquad \leq \epsilon^{1+\mu} CK(u^l, v^l) z_i^{j+1}(l) \leq (1 - \delta) z_i^{j+1}(l), \\ |w_{i+1}^j(l) - g_t(w_i^j(l), z_i^j(l), \ldots, z_{i+\beta}^j(l))| \\ \qquad \leq \epsilon^{\mu+1} |\mathbf{E}_2(\epsilon, t, u^l, v^l, \ldots)| \\ \qquad \qquad \leq \epsilon^{1+\mu} CK(u^l, v^l) w_{i+1}^j(l) \leq (1 - \delta) w_{i+1}^j(l). \end{cases}$$

In particular there is an integer $N_0 \geq \max(\delta^{-1}, 2 - \delta)$ so that the inequalities hold:

$$\begin{cases} \frac{1}{N_0} f_t^l(w_i^j(l), z_i^j(l), \ldots, z_{i+\alpha}^j(l)) \leq z_i^{j+1}(l) \leq N_0 f_t^l(w_i^j(l), z_i^j(l), \ldots, z_{i+\alpha}^j(l)) \\ \frac{1}{N_0} g_t^l(w_i^j(l), z_i^j(l), \ldots, z_{i+\beta}^j(l)) \leq w_{i+1}^j(l) \leq N_0 g_t^l(w_i^j(l), z_i^j(l), \ldots, z_{i+\beta}^j(l)). \end{cases}$$

Now $\frac{1}{N_0}f$ and $N_0 f$ are both tropically equivalent to f, with the bounds $\max(M_{\frac{1}{N_0}f}, M_{N_0 f}) \leq N_0 M_f$. Then it follows from corollary 15.1 and proposition 6.3 that the uniform estimates:

$$\left(\frac{z_i^{j+1}(1)}{z_i^{j+1}(2)}\right)^{\pm 1}, \quad \left(\frac{w_{i+1}^j(1)}{w_{i+1}^j(2)}\right)^{\pm 1}$$

$$\leq (N_0 M)^{10 P_{i+j(\gamma+1)}(c)} ([(u^1, v^1) : (u^2, v^2)]_\epsilon)^{3\tilde{c}^{i+1+j(\gamma+1)}}$$

$$\leq (N_0 M)^{10 P_{\epsilon^{-1}(x+s(\gamma+1))}(c)} ([(u^1, v^1) : (u^2, v^2)]_\epsilon)^{3\tilde{c}^{\epsilon^{-1}(x+s(\gamma+1)+1)}}$$

where $(x, s) = ((i+a)\epsilon, (j+b)\epsilon)$.

Since the right hand side does not depend on a, b, one obains the uniform estimates:

$$\left(\frac{u^1}{u^2}\right)^{\pm 1}(x, s), \quad \left(\frac{v^1}{v^2}\right)^{\pm 1}(x, s)$$

$$\leq (N_0 M)^{10 P_{\epsilon^{-1}(x+s(\gamma+1))}(c)} ([(u, v) : (u', v')]_\epsilon)^{3 c^{\epsilon^{-1}(x+s(\gamma+1)+1)}}$$

This completes the proof of theorem 15.4.

15.5 PDE systems and equivalent automata

For $l = 1, 2$, let

$$\mathbf{A}_l : \begin{cases} \psi^l : Q \times S^{\alpha+1} \to S, \\ \phi^l : Q \times S^{\beta+1} \to Q \end{cases}$$

be the equivalent automata over $R \subset Q$, and choose embeddings $Q, S \subset \mathbb{R}$.

For $l = 1, 2$, let us take their stable extensions with the constants (δ, μ):

$$\begin{cases} \psi^l : \mathbb{R} \times \mathbb{R}^{\alpha+1} \to \mathbb{R}, \\ \phi^l : \mathbb{R} \times \mathbb{R}^{\beta+1} \to \mathbb{R} \end{cases}$$

and let (f_t^l, g_t^l) and (ψ_t^l, ϕ_t^l) be the corresponding functions respectively. Then we obtain the state systems of rational dynamics:

$$\begin{cases} z_i^{j+1}(l) = f_t^l(w_i^j(l), z_i^j(l), \ldots, z_{i+\alpha}^j(l)), \\ w_{i+1}^j(l) = g_t^l(w_i^j(l), z_i^j(l), \ldots, z_{i+\beta}^j(l)). \end{cases}$$

Let us follow the process in 15.2, and induce the systems of partial differential equations of order μ:

$$\begin{cases} P_1^l(\epsilon, t, u^l, v^l, u_x^l, u_s^l, \ldots, u_{\mu s}^l) = 0 \\ P_2^l(\epsilon, t, u^l, v^l, u_x^l, v_x^l, \ldots, u_{\mu x}^l) = 0 \end{cases}$$

with the scaling parameters:

$$i = \frac{x}{\epsilon}, \quad j = \frac{s}{\epsilon}, \quad u^l(x,s) = z_i^j(l), \quad v^l(x,s) = w_i^j(l).$$

The behaviors of the orbits $\{z_i^j(l)\}_{i,j,l}$ of the rational dynamics above can be controlled by corollary 15.2, while the dynamics of the states $\{w_i^j(l)\}_{i,j,l}$ will behave quite differently from each other. One might expect parallel situation for the systems of the partial differential equations. In this case behaviors of v^1 and v^2 will still be out of control, which corresponds to the dynamics of the states.

Let us fix any positive number $0 < \tau < 0.5$, and put the domains:

$$D_1(\epsilon, \tau) = \{\, i\epsilon + a \in [0, \infty) : i \in \mathbb{N}, \ |a| \le \epsilon\tau \,\},$$
$$D(\epsilon, \tau) = D_1(\epsilon, \tau) \times D_1(\epsilon, \tau) \subset [0, \infty) \times [0, \infty).$$

$D(\epsilon)$ are the disjoint unions of the squares.

We also put the initial subdomain:

$$\begin{cases} I_x(\epsilon, \tau) = D_1(\epsilon, \tau) \times [0, \tau\epsilon) \subset I_x = [0, \infty) \times [0, \tau\epsilon), \\ I_s(\epsilon, \tau) = [0, \tau\epsilon) \times D_1(\epsilon, \tau) \subset I_s = [0, \tau\epsilon) \times [0, \infty). \end{cases}$$

Theorem 15.5. *Let C_0 be the error constants. Let $u^l, v^l : (0, \infty) \times [0, \infty) \to (0, \infty)$ be solutions to the above systems respectively. Then for any $C \ge 1$, there exists $t_0 > 1$ and $D \ge 1$ so that the following holds for all $t \ge t_0$:*
Suppose two conditions: (1) The estimates:

$$0 \le CK(u^l, v^l) \le (1 - \delta)\epsilon^{-1-\mu}$$

are satisfied for some positive $\delta > 0$.
(2) The inclusions hold:

$$u^l|I_x(\epsilon, \tau) \subset N_C(t^S), \quad v^l|I_s(\epsilon, \tau) \subset N_C(t^Q).$$

Then they satisfy the uniform bounds:

$$\left(\frac{u^1}{u^2}\right)^{\pm 1}(x, s) \le D$$

for all $(x, y) \in D(\epsilon, \tau)$.

The proof is given in sec. 15.6.

15.6   Dynamical inequalities under change of automata

For $l = 1, 2$, let:

$$
\mathbf{A}_l : \left\{ \begin{array}{l} \psi^l : Q \times S^{\alpha+1} \to S, \\ \phi^l : Q \times S^{\beta+1} \to Q \end{array} \right.
$$

be the automata which are equivalent over $R \subset Q$.

For each l, let us consider two pairs of their stable extensions with the constants (δ, μ) by relative $(\max, +)$-functions for $m = 1, 2$:

$$
\left\{ \begin{array}{l} \psi^{l,m} : \mathbb{R} \times \mathbb{R}^{\alpha+1} \to \mathbb{R}, \\ \phi^{l,m} : \mathbb{R} \times \mathbb{R}^{\beta+1} \to \mathbb{R}. \end{array} \right.
$$

Suppose that for each $l = 1, 2$:

$$
(\psi^{l,1}, \phi^{l,1}) \sim (\psi^{l,2}, \phi^{l,2})
$$

are pairwisely equivalent. Let $(f_t^{l,m}, g_t^{l,m})$ and $(\psi_t^{l,m}, \phi_t^{l,m})$ be their tropical correspondences to $(\psi^{l,m}, \phi^{l,m})$ respectively.

Corollary 15.2. *Assume the above conditions. Then for any large $C \geq 1$, there exists $t_0 > 1$ and $D \geq 1$ so that the following holds for all $t \geq t_0$:*

Suppose two sequences $\{w_i^j(l)\}_{i,j}$ and $\{z_i^j(l)\}_{i,j}$ satisfy the dynamical inequalities for $l = 1, 2$:

$$
\left\{ \begin{array}{l} f_t^{l,1}(w_i^j(l), z_i^j(l), \dots, z_{i+\alpha}^j(l)) \leq z_i^{j+1}(l) \leq f_t^{l,2}(w_i^j(l), z_i^j(l), \dots, z_{i+\alpha}^j(l)), \\ g_t^{l,1}(w_i^j(l), z_i^j(l), \dots, z_{i+\beta}^j(l)) \leq w_{i+1}^j(l) \leq g_t^{l,2}(w_i^j(l), z_i^j(l), \dots, z_{i+\beta}^j(l)). \end{array} \right.
$$

Moreover suppose the initial values are contained in the same C neighborhoods:

$$
\bar{z}(l) = (z_0(l), z_1(l), \dots) \in N_C(t^{\bar{k}}), \quad \bar{w}(l) = (w^0(l), w^1(l), \dots) \in N_C(t^{\bar{q}})
$$

for some $\bar{k} = (k_0, k_1, \dots) \in X_S$ and $\bar{q} = (q^0, q^1, \dots) \in X_R$.

Then the uniform estimates hold:

$$
\max \left\{ \frac{z_i^j(1))}{z_i^j(2)}, \ \frac{z_i^j(2)}{z_i^j(1)} \right\} \leq D.
$$

Proof. The proof is long and we split it into several steps, but the idea is quite parallel to theorem 8.1 and corollary 15.1. Notice that the dynamics of the states $w_i^j(1)$ and $w_i^j(2)$ can be very different from each other, and the distribution of their ratios will be out of control in general.

Step 1: Let us fix large $t \gg 1$. By replacing δ by a smaller one as in Step 1 in the proof of theorem 8.1, we may assume the following:

(1) $\delta = \log_t C$ and the estimate:

$$M^2 < C^{1-\mu}$$

holds. In particular the estimate $2\log_t M + \mu\delta < \delta$ holds.

(2) $(\psi^{l,m}, \phi^{l,m})$ are all $(3\delta, \mu)$- stable.

Let $p_i^j(l) = \log_t z_i^j(l)$ and $o_i^j(l) = \log_t w_i^j(l)$. Then we have the estimates:

$$\begin{cases} \psi_t^{l,1}(o_i^j(l), p_i^j(l), \dots, p_{i+\alpha}^j(l)) \leq p_i^{j+1}(l) \leq \psi_t^{l,2}(o_i^j(l), p_i^j(l), \dots, p_{i+\alpha}^j(l)), \\ \phi_t^{l,1}(o_i^j(l), p_i^j(l), \dots, p_{i+\beta}^j(l)) \leq o_{i+1}^j(l) \leq \phi_t^{l,2}(o_i^j(l), p_i^j(l),, \dots, p_{i+\beta}^j(l)) \end{cases}$$

where $p_i^0(l) = x_i(l) \equiv \log_t z_i(l)$ and $o_0^j(l) = y^j(l) \equiv \log_t w^j(l)$.

Notice that by assumption, both the equalities:

$$\begin{cases} \psi^{l,1} = \psi^{l,2}, \\ \phi^{l,1} = \phi^{l,2} \end{cases}$$

hold as functions.

We will use the notation ψ^l and ϕ^l otherwise stated. Let us consider another solution to the state systems:

$$\begin{cases} k_i^{j+1}(l) = \psi^l(q_i^j(l), k_i^j(l), \dots, k_{i+\alpha}^j(l)) \\ q_{i+1}^j(l) = \phi^l(q_i^j(l), k_i^j(l),, \dots, k_{i+\beta}^j(l)) \end{cases}$$

with $k_i^0(l) = k_i$ and $q_0^j(l) = q^j$.

Let us verify the uniform estimates:

$$|p_i^j(l) - k_i^j(l)|, \quad |o_i^j(l) - q_i^j(l)| < \delta$$

by several steps below.

Step 2: Let us introduce another state system:

$$\begin{cases} x'_{l,m}(i, j+1) = \psi_t^{l,m}(y'_{l,m}(i,j), x'_{l,m}(i,j), \dots, x'_{l,m}(i+\alpha, j)), \\ y'_{l,m}(i+1, j) = \phi_t^{l,m}(y'_{l,m}(i,j), x'_{l,m}(i,j), \dots, x'_{l,m}(i+\beta, j)), \end{cases}$$

$$\begin{cases} x_l(i, j+1) = \psi^l(y_l(i,j), x_l(i,j), \dots, x_l(i+\alpha, j)), \\ y_l(i+1, j) = \phi^l(y_l(i,j), x_l(i,j), \dots, x_l(i+\beta, j)), \end{cases}$$

with the same initial values $x'_{l,m}(i, 0) = x_l(i, 0) = x_i(l)$ and $y'_{l,m}(0, j) = y_l(0, j) = y^j(l)$.

By lemma 8.2, the estimates hold:
$$|x_l(i,j) - k_i^j(l)|, \ |y_l(i,j) - q_i^j(l)| < \delta.$$
On the other hand by lemma 8.3, other estimates hold:
$$|x_l(i,j) - x'_{l,m}(i,j)|, \quad |y_l(i,j) - y'_{l,m}(i,j)| < \delta.$$
So combining these estimates, one obtains the estimates:
$$|x'_{l,m}(i,j) - k_i^j(l)|, \quad |y'_{l,m}(i,j) - q_i^j(l)| < 2\delta.$$

Step 3: Next let us verify the uniform estimates for $l = 1, 2$:
$$|x'_{l,m}(i,1) - p_i^1(l)|, \ |y'_{l,m}(i,0) - \sigma_i^0(l)| \ < \delta.$$
Notice the equalities $y'_{l,m}(0,0) = \sigma_0^0(l)$ and $x'_{l,m}(i,0) = p_i^0(l)$.

Let us verify $|y'_{l,m}(i,0) - \sigma_i^0(l)| \ < \delta$ by induction on $i = 0, 1, 2, \ldots$
Assume that the estimates hold up to i.

Firstly suppose $\sigma_{i+1}^0(l) \geq y'_{l,1}(i+1,0)$ holds. Then the estimates hold:
$$0 \leq |\sigma_{i+1}^0(l) - y'_{l,1}(i+1,0)| = \sigma_{i+1}^0(l) - y'_{l,1}(i+1,0)$$
$$\leq \phi_t^{l,2}(\sigma_i^0(l), p_i^0(l), \ldots, p_{i+\beta}^0(l)) - \phi_t^{l,1}(y'_{l,1}(i,0), x'_{l,1}(i,0), \ldots, x'_{l,1}(i+\beta,0))$$
$$\leq 2\log_t M +$$
$$|\phi^l(\sigma_i^0(l), p_i^0(l), \ldots, p_{i+\beta}^0(l)) - \phi^l(y'_{l,1}(i,0), x'_{l,1}(i,0), \ldots, x'_{l,1}(i+\beta,0))|$$
$$\leq 2\log_t M + \mu|\sigma_i^0(l) - y'_{l,1}(i,0)| \leq 2\log_t M + \mu\delta < \delta.$$
Conversely suppose $\sigma_{i+1}^0(l) \leq y'_{l,1}(i+1,0)$ holds. Then the estimates hold:
$$0 \leq |y'_{l,1}(i+1,0) - \sigma_{i+1}^0(l)| = y'_{l,1}(i+1,0) - \sigma_{i+1}^0(l)$$
$$\leq \phi_t^{l,1}(y'_{l,1}(i,0), x'_{l,1}(i,0), \ldots, x'_{l,1}(i+\beta,0)) - \phi_t^{l,1}(\sigma_i^0(l), p_i^j(l), \ldots, p_{i+\beta}^0(l))$$
$$\leq 2\log_t M +$$
$$|\phi^l(y'_{l,1}(i,0), x'_{l,1}(i,0), \ldots, x'_{l,1}(i+\beta,0)) - \phi^l(\sigma_i^0(l), p_i^0(l), \ldots, p_{i+\beta}^0(l))|$$
$$\leq 2\log_t M + \mu|\sigma_i^0(l) - y'_{l,1}(i,0)| \leq 2\log_t M + \mu\delta < \delta.$$
Thus in all cases, the estimate $|y'_{l,1}(i+1,0) - \sigma_{i+1}^0(l)| \ < \delta$ holds. By the induction step, we have verified the claim. In the same way the estimates $|y'_{l,2}(i,0) - \sigma_i^0(l)| \ < \delta$ also hold for all i.

Then using the above estimates, we follow a parallel argument as below. Suppose $p_i^1(l) \geq x'_{l,1}(i,1)$ holds for some i. Then we have the estimates:
$$0 \leq |p_i^1(l) - x'_{l,1}(i,1)| = p_i^1(l) - x'_{l,1}(i,1)$$
$$\leq \psi_t^{l,2}(\sigma_i^0(l), p_i^0(l), \ldots, p_{i+\alpha}^0(l)) - \psi_t^{l,1}(y'_{l,1}(i,0), x'_{l,1}(i,0), \ldots, x'_{l,1}(i+\beta,0))$$
$$\leq 2\log_t M +$$
$$|\psi^l(\sigma_i^0(l), p_i^0(l), \ldots, p_{i+\alpha}^0(l)) - \psi^l(y'_{l,1}(i,0), x'_{l,1}(i,0), \ldots, x'_{l,1}(i+\beta,0))|$$
$$\leq 2\log_t M + \mu|\sigma_i^0(l) - y'_{l,1}(i,0)| \leq 2\log_t M + \mu\delta < \delta.$$

Conversely suppose $p_i^j(l) \leq x_{l,1}'(i,1)$ holds. Then we have the estimates:

$$0 \leq |x_{l,1}'(i,1) - p_i^1(l)| = x_{l,1}'(i,1) - p_i^1(l)$$

$$\leq \psi_t^{l,1}(y_{l,1}'(i,0), x_{l,1}'(i,0), \ldots, x_{l,1}'(i+\beta,0)) - \psi_t^{l,1}(\sigma_i^0(l), p_i^0(l), \ldots, p_{i+\alpha}^0(l))$$

$$\leq 2\log_t M +$$

$$|\psi^l(y_{l,1}'(i,0), x_{l,1}'(i,0), \ldots, x_{l,1}'(i+\beta,0)) - \psi^l(\sigma_i^0(l), p_i^0(l), \ldots, p_{i+\alpha}^0(l))|$$

$$\leq 2\log_t M + \mu|\sigma_i^0(l) - y_{l,1}'(i,0)| \leq 2\log_t M + \mu\delta < \delta.$$

Thus in any cases we have the estimates $|x_{l,1}'(i,1) - p_i^1(l)| < \delta$ for all $i = 0, 1, 2, \ldots$ The estimates $|x_{l,2}'(i,1) - p_i^1(l)| < \delta$ are obtained in the same way.

Step 4: Let us verify the estimates for all $i, j \geq 0$:

$$|x_{l,m}'(i,j) - p_i^j(l)|, \ |y_{l,m}'(i,j) - \sigma_i^j(l)| \ < \delta.$$

Let us put the sequences:

$$\bar{x}_{l,m}'(j) = (x_{l,m}'(0,j), x_{l,m}'(1,j), x_{l,m}'(2,j), \ldots)$$

and similarly for $\bar{y}_{l,m}'(j)$, $\bar{p}^j(l)$ and $\bar{\sigma}^j(l)$.

Let us verify the estimates:

$$d(\bar{x}_{l,m}'(j+1), \bar{p}^{j+1}(l)), \ d(\bar{y}_{l,m}'(j), \bar{\sigma}^j(l)) \ < \delta$$

by induction on $j = 0, 1, 2, \ldots$ We have verified the estimates for $j = 0$ at step 3. So assume they hold up to $j - 1$.

The initial conditions $\sigma_0^j(l) = y_{l,1}'(0,j)$ hold. Let us verify the estimates $|\sigma_i^j(l) - y_{l,1}'(i,j)| < \delta$ by induction on i. Suppose they hold up to i.

Firstly suppose $\sigma_{i+1}^j(l) \geq y_{l,1}'(i+1,j)$ holds. Then the estimates hold:

$$0 \leq |\sigma_{i+1}^j(l) - y_{l,1}'(i+1,j)| = \sigma_{i+1}^j(l) - y_{l,1}'(i+1,j)$$

$$\leq \phi_t^{l,2}(\sigma_i^j(l), p_i^j(l), \ldots, p_{i+\beta}^j(l)) - \phi_t^{l,1}(y_{l,1}'(i,j), x_{l,1}'(i,j), \ldots, x_{l,1}'(i+\beta,j))$$

$$\leq 2\log_t M +$$

$$|\phi^l(\sigma_i^j(l), p_i^j(l), \ldots, p_{i+\beta}^j(l)) - \phi^l(y_{l,1}'(i,j), x_{l,1}'(i,j), \ldots, x_{l,1}'(i+\beta,j))|$$

$$\leq 2\log_t M +$$

$$\mu \max(|\sigma_i^j(l) - y_{l,1}'(i,j)|, |p_i^j(l) - x_{l,1}'(i,j)|, \ldots, |p_{i+\beta}^j(l) - x_{l,1}'(i+\beta,j)|)$$

$$\leq 2\log_t M + \mu\delta < \delta.$$

The converse case can be estimated in the same way as step 3, and we omit repetition. So we have the estimates $|y_{l,1}'(i,j) - \sigma_i^j(l)| < \delta$ by the induction step for all i. In the same way we can verify the estimates $|y'_{l,2}(i,j) - \sigma_i^j(l)| < \delta$.

Then we have the estimates $|x'_{l,m}(i, j+1) - p_i^{j+1}(l)| < \delta$ for all $i = 0, 1, 2, \ldots$ again in the same way as step 3.

This completes the induction step on j, and so we have obtained the desired estimates:

$$|x'_{l,m}(i,j) - p_i^j(l)|, \ |y'_{l,m}(i,j) - \sigma_i^j(l)| \ < \delta$$

for all $i, j = 0, 1, 2, \ldots$

Step 5: Combining steps 2 $\sim$ 4, we obtain the estimates:

$$|p_i^j(l) - k_i^j(l)| < 3\delta.$$

In particular all the values of these sequences lie within μ-Lipschitz constants of $\psi^{l,m}$ and $\phi^{l,m}$.

Let us consider the state systems of the rational dynamics:

$$\begin{cases} F_i^{j+1}(l) = f_t^{l,1}(G_i^j(l), F_i^j(l), \ldots, F_{i+\alpha}^j(l)) \\ G_i^{j+1}(l) = g_t^{l,1}(G_i^j(l), F_i^j(l), \ldots, F_{i+\beta}^j(l)) \end{cases}$$

with the initial values $F_i^0(l) = z_i(l)$ and $G_0^j(l) = w^j(l)$ respectively. Then we apply proposition 15.1, and obtain the uniform estimates:

$$\max\left(\frac{F_i^{j+1}(l)}{z_i^{j+1}(l)}, \frac{z_i^{j+1}(l)}{F_i^{j+1}(l)}\right) \leq M^{2P_{i+j(\gamma+1)}(\mu)} \leq C' \qquad (l = 1, 2)$$

for some C', since $P_i(\mu)$ are uniformly bounded for $0 < \mu < 1$. On the other hand by theorem 8.1, the estimates hold:

$$\max\left(\frac{F_i^{j+1}(1)}{F_i^{j+1}(2)}, \frac{F_i^{j+1}(2)}{F_i^{j+1}(1)}\right) \leq C^4.$$

Thus combining these, one obtains the desired estimates:

$$\left(\frac{z_i^j(1)}{z_i^j(2)}\right)^{\pm 1} = \left(\frac{z_i^j(1)}{F_i^j(1)}\right)^{\pm 1}\left(\frac{F_i^j(2)}{z_i^j(2)}\right)^{\pm 1} \leq \ C' C^4 \equiv D.$$

$\square$

Proof of theorem 15.5.

We follow a similar argument as the proof of theorem 15.4. Let $N_0 \geq \max(\delta^{-1}, 2 - \delta)$ be an integer.

Let us fix $0 \le a, b \le \tau$, and put the subdomain lattice as:

$$D_{\epsilon,\tau}(a,b) = \{(\,(l_1+a)\epsilon, (l_2+b)\epsilon\,) \in [0,\infty) \times [0,\infty) : l_1, l_2 \in \mathbb{N}\}.$$

Let the pairs of the orbits which are given by the values at $D_{\epsilon,\tau}(a,b)$ be:

$$\begin{cases} z_i^j(l) = u^l((i+a)\epsilon, (j+b)\epsilon), \\ w_i^j(l) = v^l((i+a)\epsilon, (j+b)\epsilon). \end{cases}$$

Then by the same way as the proof of theorem 15.4, we obtain the inequalities:

$$\begin{cases} \frac{1}{N_0} f_t^l(w_i^j(l), z_i^j(l), \dots, z_{i+\alpha}^j(l)) \le z_i^{j+1}(l) \le N_0 f_t^l(w_i^j(l), z_i^j(l), \dots, z_{i+\alpha}^j(l)), \\ \frac{1}{N_0} g_t^l(w_i^j(l), z_i^j(l), \dots, z_{i+\beta}^j(l)) \le w_{i+1}^j(l) \le N_0 g_t^l(w_i^j(l), z_i^j(l), \dots, z_{i+\beta}^j(l)). \end{cases}$$

Notice that $\frac{1}{N_0} f$ and $N_0 f$ are both tropically equivalent to f.

By assumption, the initial values lie on the subdomain:

$$\begin{cases} z_i^0(l) = u^l((i+a)\epsilon, b\epsilon) \in N_C(t^S), \\ w_0^j(l) = v^l(a\epsilon, (j+b)\epsilon) \in N_C(t^R). \end{cases}$$

It follows from corollary 15.2 that the uniform estimates hold:

$$\max\left(\frac{z_i^j(1)}{z_i^j(2)}, \frac{z_i^j(2)}{z_i^j(1)}\right) \le D.$$

Since the right hand side does not depend on a, b, one obains the uniform bounds:

$$\left(\frac{u^1}{u^2}\right)^{\pm 1}(x, s) \le D$$

for all $(x,y) \in D(\epsilon, \tau)$.

This completes the proof of theorem 15.5.

References

This chapter gave a comparison analysis of global solutions to different hyperbolic Mealy systems in [Kat9]. More concrete instances of hyperbolic equations are given in [Log], and more general classes of hyperbolic PDE systems are introduced in [BGH].

Chapter 16

Analysis of hyperbolic Mealy systems

We develop a basic analysis of the hyperbolic Mealy systems of partial differential equations in our sense. In particular we verify the existence and uniqueness of their global solutions. Then we apply the previous results to the large scale analysis of them.

16.1 Hyperbolic Mealy systems

Let us study the basic analysis of the 1st order hyperbolic systems of partial differential equations with 2 variables.

Let (f_t, g_t) be a pair of elementary rational functions, corresponding to the pair of relative (max, +)-functions (ψ, ϕ).

Let us consider the corresponding hyperbolic Mealy systems:

$$\begin{cases} \epsilon\, u_s = f_t(v, u) - u, \\ \epsilon\, v_x = g_t(v, u) - v. \end{cases}$$

Recall the higher distortion for the Mealy systems:

$$K(u, v) \equiv \sup_{(x,s)\in[0,\infty)^2} \max \Big[\ \frac{||(u, v)||^1_{1,0}}{u(x, s + \epsilon)},\ \ \frac{||(u, v)||^2_{1,0}}{v(x + \epsilon, s)}\ \Big].$$

Definition 16.1. The pair (f_t, g_t) is admissible of Mealy type, if there are solutions $u, v : [0, \infty) \times [0, \infty) \to [0, \infty)$ which satisfy the estimates:

$$\{\ |(f_t(v, u) - u)((f_t)_u(v, u) - 1)| + \epsilon|v_s(f_t)_v(u, v)|\ \}(x, s)$$
$$< (2 - \mu)u(x, s + \alpha),$$

$$\{\ |(g_t(v, u) - v)((g_t)_u(v, u) - 1)| + \epsilon|u_x(g_t)_v(u, v)|\ \}(x, s)$$
$$< (2 - \mu)v(x + \alpha, s).$$

for all $0 \le \alpha \le 1$ and $(x, s) \in [0, \infty) \times [0, \infty)$.

Corollary 16.1. *For $l = 1, 2$, let (f_t^l, g_t^l) be admissible pairs, and (u^l, v^l) :*
$[0, \infty) \times [0, \infty) \to (0, \infty)$ be the solutions which satisfy the above conditions.
Then they satisfy the asymptotic estimates:

$$(\frac{u^1}{u^2})^{\pm 1}(x, s), \quad (\frac{v^1}{v^2})^{\pm 1}(x, s)$$

$$\leq (M_0)^{10 P_{\epsilon^{-1}(x+s(\gamma+1))}(c)} \ ([(u^1, v^1) : (u^2, v^2)]_\epsilon)^{3\tilde{c}^{-1}(x+s(\gamma+1)+1)}$$

for some M_0 and all $(x, s) \in [0, \infty) \times [0, \infty)$.

Proof. For a Mealy system, the error constant is always $\frac{1}{2}$. Moreover we have the equalities:

$$\begin{cases} \epsilon^2 \frac{\partial^2 u}{\partial s^2} = (f_t(v, u) - u)((f_t)_u(v, u) - 1) + (f_t)_v(u, v)\epsilon v_s, \\ \epsilon^2 \frac{\partial^2 v}{\partial x^2} = (g_t(v, u) - v)((g_t)_v(v, u) - 1) + (g_t)_u(u, v)\epsilon u_x. \end{cases}$$

So in order to apply theorem 15.4, the required conditions on the higher distortion are the uniform bounds:

$$\frac{|(f_t(v, u) - u)((f_t)_u(v, u) - 1)| + \epsilon |v_s(f_t)_v(u, v)|(x, s)}{2u(x, s + \alpha)} < (1 - \delta),$$

$$\frac{|(g_t(v, u) - v)((g_t)_v(v, u) - 1)| + \epsilon |u_x(g_t)_u(u, v)|(x, s)}{2v(x + \alpha, s)} < (1 - \delta)$$

for some $0 < \delta < 1$ and $0 < \epsilon \leq 1$, which follow from admissibility of Mealy type. $\qquad \square$

16.2   Refinement

Let:

$$\mathbf{A} : \begin{cases} \psi : Q \times S \to S, \\ \phi : Q \times S \to Q \end{cases}$$

be a Mealy automaton. For an initial value $\bar{q} = (q^0, q^1, \dots) \in X_Q$ and $\bar{k} = (k_0, k_1, \dots) \in X_S$, let us denote the orbits by $\{k_i^j\}$ and $\{q_i^j\}$.
Let:

$$\begin{cases} \tilde{\psi} : \mathbb{R} \times \mathbb{R} \to \mathbb{R}, \\ \tilde{\phi} : \mathbb{R} \times \mathbb{R} \to \mathbb{R} \end{cases}$$

be two maps. Suppose for some $a, b \in \mathbb{R} \cup \{\pm\infty\}$, the restriction of the pair of the maps has their range which is included in the domain:

$$(\tilde{\psi}, \tilde{\phi}) : [a, b]^2 \to [a, b]^2.$$

We say that the pair restricts to a self-map over $[a, b]^2$.

Let us denote:

$$X_{[a,b]} = \{(x_0, x_1, \dots); \ x_i \in [a, b]\}$$

and take an initial value $\bar{x} = (x_0, x_1, \dots), \bar{y} = (y^0, y^1, \dots) \in X_{[a,b]}$. Then we have the orbits $\{x_i^j\}$ and $\{y_i^j\}$ given by the system of the equations:

$$\begin{cases} x_i^{j+1} = \tilde{\psi}(y_i^j, x_i^j), \\ y_{i+1}^j = \tilde{\phi}(y_i^j, x_i^j). \end{cases}$$

Let us choose embeddings $S, Q \subset [a, b]$.

Definition 16.2. $(\tilde{\psi}, \tilde{\phi})$ is an ϵ-refinement of the pair (ψ, ϕ), if there is a posiitive number N so that for any $\bar{q} \in X_Q$ and $\bar{k} \in X_S$, there are $\bar{x} = (x_0, x_1, \dots), \bar{y} = (y^0, y^1, \dots) \in X_{\mathbb{R}}$ with:

$$\begin{cases} y^{jN} = q^j, & |y^{j+1} - y^j| \leq \epsilon, \\ x_{iN} = k_i, & |x_{i+1} - x_i| \leq \epsilon \end{cases}$$

for all $i, j \in \{0, 1, \dots\}$, such that the equalities hold:

$$\begin{cases} x_{iN}^{jN} = k_i^j, \\ y_{iN}^{jN} = q_i^j \end{cases}$$

where $\{(x_i^j, y_i^j)\}_{i,j}$ is the orbit as above.

(ψ, ϕ) is refinable, if there is an ϵ-refinement for any small $\epsilon > 0$.

An ϵ refinement is almost diagonal, if moreover they satisfy the estimates for all $(x, y) \in [a, b]$:

$$|(x, y) - (\tilde{\psi}(x, y), \tilde{\phi}(x, y))| \leq \epsilon.$$

Later we verify the existence of admissible solutions by use of refinement and their exponential asymptotic estimates.

16.3   Existence of admissible solutions

Let **A** be a Mealy automaton with 2 alphabets. By proposition 8.1, there exists a stable extension (ψ, ϕ) of **A**.

For its refinement $(\tilde{\psi}, \tilde{\phi})$ and their tropical correspondences $(\tilde{f}_t, \tilde{g}_t)$, let us consider the hyperbolic Mealy systems:

$$\begin{cases} u_s = \tilde{f}_t(v, u) - u, \\ v_x = \tilde{g}_t(v, u) - v. \end{cases}$$

Theorem 16.1. *There is a refinement $(\tilde{\psi}, \tilde{\phi})$ of (ψ, ϕ) with the pair of tropical correspondences $(\tilde{f}_t, \tilde{g}_t)$ so that the hyperbolic Mealy system given by $(\tilde{f}_t, \tilde{g}_t)$ admits admissible solutions.*

The proof consists of two things: where one is to verify the existence of global solutions, and the other is to construct an admissible refinement.

These follow from theorem 16.2 and proposition 16.3 below.

It would be of interest to apply group-theoretic results to the analysis of PDEs. In particular concerning the dynamical Burnside problem in chapter 9, we would like to propose the following.

Let $X \subset [0, \infty)$ be a subspace, and denote $C^\infty(\mathbb{N}, X) \subset C^\infty[0, \infty)$ which consists of smooth functions with $f(i) \in X$ for all $i \in \mathbb{N}$.

Conjecture 16.1. *There exists a Mealy automaton and its stable extension (f_t, g_t) such that the following holds: there exist pairs of parametrized subspaces (X_t, Y_t) for $t > 1$ and solutions (u_t, v_t) to the hyperbolic Mealy system on (f_t, g_t) with their initial values $u(x, 0) \in C^\infty(\mathbb{N}, X_t)$ and $v(0, s) \in C^\infty(\mathbb{N}, Y_t)$ such that for $i, j \in \mathbb{N}$, the values:*

$$z_i^j(t) = u_t(i, j)$$

consist of infinitely quasi-recursive orbits in the sense of definition 9.1.

Remark 16.1. Once the existence of solutions is guaranteed, one can try to transform the characteristic properties of automata semi-groups to solutions to the corresponding hyperbolic Mealy systems, which in the case of groups, include amenability, random walks and degree of the growth functions.

16.4　On ϵ dependence

Let $(u, v) : [0, \infty) \times [0, \infty) \to (0, \infty)^2$ be a solution to the hyperbolic Mealy system:

$$
\begin{cases}
\epsilon\, u_s = f_t(v, u) - u, \\
\epsilon\, v_x = g_t(v, u) - v
\end{cases}
$$

with finite higher distortion $K(u, v) < \infty$.

In order to apply the asymptotic comparison for solutions to the systems of the partial differential equations, their solutions are required to satisfy some bounds:

$$
CK(u, v) \leq (1 - \delta)\epsilon^{-2}
$$

for some $0 < \delta < 1$, where C is the error constant.

For small $0 < \tau$, let us consider the reparametrized pair of the functions:

$$
(\tilde{u}, \tilde{v})(x, s) \equiv (u, v)(\tau x, \tau s)
$$

which satisfy the equations:

$$
\begin{cases}
\epsilon\tau^{-1}\, \tilde{u}_s = f_t(\tilde{v}, \tilde{u}) - \tilde{u}, \\
\epsilon\tau^{-1}\, \tilde{v}_x = g_t(\tilde{v}, \tilde{u}) - \tilde{v}.
\end{cases}
$$

Notice that the higher distortions are related as:

$$
\tau^2 K_2(u, v) = K_2(\tilde{u}, \tilde{v}).
$$

The required condition with respect to the reparametrized pair is given by:

$$
CK_2(\tilde{u}, \tilde{v}) \leq (1 - \delta)\tau^2\epsilon^{-2}.
$$

So this reparametrization is not effective for our later purpose. Later on we will put:

$$
\epsilon = 1.
$$

16.5　Basic PDE analysis, existence and uniqueness

Let us start from some analytic properties of hyperbolic Mealy systems. Let (ψ, ϕ) be a pair of relative $(\max, +)$-functions, and (f_t, g_t) be the corresponding pair of the elementary rational functions.

Lemma 16.1. *(1) Suppose the pair* (ψ, ϕ) *takes bounded range both from below and above. Then the corresponding pair* (f_t, g_t) *also satisfies the same property for each* $t > 1$.

Let us fix $t > 1$, *and* $(u, v) : [0, \infty) \times [0, \infty) \to \mathbb{R}^2$ *be a solution to the hyperbolic Mealy system with respect to* (f_t, g_t) *such that the initial value has positive range:*

$$u(x, 0), v(0, s) \in (0, \infty).$$

(2) Suppose that there are constants $r < R$ *so that* f_t *and* g_t *both satisfy the uniform bounds:*

$$r \le f_t(a, b), \ g_t(a, b) \le R.$$

Then the solution also has positive and bounded range:

$$u, v : [0, \infty) \times [0, \infty) \to (0, \infty).$$

(3) If f_t *and* g_t *both satisfy the bounds:*

$$r \le f_t(a, b), \ g_t(b, a) \le R$$

for any $r \le b \le R$ *and all* $a > 0$, *and if we choose the initial values with their ranges as:*

$$r \le u(x, 0), v(0, s) \le R$$

then their values are also contained in the range:

$$r \le u(x, s), v(x, s) \le R$$

for all $(x, s) \in [0, \infty) \times [0, \infty)$.

Proof. (1) follows from proposition 2.1 and lemma 3.4.

Let us fix $x \in [0, \infty)$ and $v(x, s)$. Then we let $u(x, \ ) : [0, \infty) \to \mathbb{R}$ satisfy the ordinary differential equation on the s variable:

$$\epsilon u_s(x, \ ) = f_t(v(x, s), u(x, \ )) - u(x, \ ).$$

For (2) and (3), because the range of f_t is away from 0, if $u(x, \ )$ takes a small value at some s, then $u_s(x, s)$ becomes positive and so their values must increase on the s variable. Conversely if $u(x, \ )$ takes large values, then $u_s(x, \ )$ becomes negative and their values must decrease.

For the $v(\ , s)$ case, we can argue in the same way.   □

Definition 16.3. The pair (f_t, g_t) gives a stable Mealy dynamics over $[r, R]$ (for a fixed $t > 1$), if there is some $0 < q < r$ so that they satisfy the bounds for all $a > 0$:

$$f_t(a, b) - b, \quad g_t(b, a) - b \quad \begin{cases} \geq q & b \leq r + q \\ \leq -q & b \geq R - q \end{cases}$$

$$|f_t(a, b) - b|, \quad |g_t(b, a) - b| \leq R - r \qquad r \leq b \leq R.$$

The following observations follow immediately by lemma 16.1:

Lemma 16.2. *(1) If (f_t, g_t) satisfies the uniform bounds as $r \leq f_t, g_t \leq R$, then the pair gives a stable Mealy dynamics over $[r - 2q, R + 2q]$.*

(2) If initial values admit the bounds $r + q \leq u(x, 0), v(0, s) \leq R - q$, then the solutions are also contained in the range:

$$r + q \leq u(x, s), v(x, s) \leq R - q$$

for all $(x, s) \in [0, \infty) \times [0, \infty)$.

16.6   Existence and uniqueness

Let us study the existence and uniqueness of solutions to the hyperbolic Mealy systems. We use an automata version of the Picard iteration method of successive approximation.

Let (f_t, g_t) be a pair of elementary rational functions of 2 variables, which corresponds to the relative $(\max, +)$-functions (ψ, ϕ). Let us assume that the pair gives a stable Mealy dynamics over $[r, R]$ with $0 < q < r$.

Let us introduce the following quantity:

$$\begin{aligned} D &= D_{r,R}(f_t, g_t) \\ &= \sup_{(v,u)\in[r,R]^2} \{|f_t(v, u) - u|, \ |g_t(v, u) - v|\} \leq R - r. \end{aligned}$$

Let $\bar{f}_t(v, u) = f_t(v, u) - u$ and $\bar{g}_t(v, u) = g_t(v, u) - v$ so that the hyperbolic Mealy equations can be written as:

$$\begin{cases} u_s = \bar{f}_t(v, u), \\ v_x = \bar{g}_t(v, u). \end{cases}$$

Now let us fix $t > 1$ and give the initial values:

$$\begin{cases} u : [0, \infty) \times \{0\} \to [r + q, R - q] \\ v : \{0\} \times [0, \infty) \to [r + q, R - q]. \end{cases}$$

With (u, v) above, let us follow the processes (1), (2), (3), (4) below to solve the equations of the hyperbolic Mealy system.

Construction of solutions:

(1) Let us choose small $\tau > 0$ so that:

(a) the Lipschitz constants of $\bar{f}_t$ and $\bar{g}_t$ are both less than $(2\tau)^{-1}$ on $[r, R]^2$.

(b) the estimate $\tau \leq D^{-1} q$ holds.

(2) For $m = 0, 1, 2, \ldots$, let us put the successive squares:

$$D_m = [m\tau, (m + 1)\tau] \times [0, \tau] \subset [0, \infty) \times [0, \infty).$$

Construct a solution by extending it on each D_m inductively.

Suppose a solution is given over D_{m-1}. With the original initial values, we have determined the values of the solution as:

$$\begin{cases} u \mid D_{m-1} \cup [0, \infty) \times \{0\}, \\ v \mid D_{m-1} \cup \{0\} \times [0, \infty). \end{cases}$$

(3) On the domain D_m, we have the initial value as:

$$\begin{cases} u \mid [x_0, x_0 + \tau] \times \{0\} \cup \{x_0\} \times [0, \tau], \\ v \mid \{x_0\} \times [0, \tau] \end{cases}$$

and let us extend the solutions over D_m.

For $(x, s) \in D_m$, let:

$$(u_0, v_0) = (u(x, 0), v(x_0, s))$$

and define the sequence inductively by:

$$(u_n, v_n) = (u_0, v_0) + \left(\int_0^s \bar{f}_t(v_{n-1}, u_{n-1}) dt, \int_{x_0}^x \bar{g}_t(v_{n-1}, u_{n-1}) dy \right).$$

Lemma 16.3. *Suppose that the pair (f_t, g_t) gives a stable Mealy dynamics over $[r, R]$. Then the sequence $\{(u_n, v_n)\}_n$ converges uniformly on D_m:*

$$(u, v) = \lim_{n \to \infty} (u_n, v_n)$$

whose limit gives the solution and coincides with the given initial value. Moreover they satisfy the estimates:

$$|(u, v) - (u_n, v_n)| \leq \frac{1}{2^{n-1}} \max(|u_1 - u_0|, |v_1 - v_0|).$$

Proof. Step 1: We claim that the ranges of u_n and v_n are both in the region $[r, R]$. We verify it only for u_n. v_n can be considered similarly.

By assumption, the uniform bounds $r + q \leq u_0 \leq R - q$ hold. Firstly we have the estimates:

$$|u_1 - u_0| \leq \int_0^\tau |f_t(v_0, u_0) - u_0| \leq \tau D \leq D^{-1} q D = q.$$

So u_1 admits the bound $r \leq u_1 \leq R$.

Let us consider the equation:

$$u_n = u_0 + \int_0^s [f_t(v_{n-1}, u_{n-1}) - u_{n-1}] dt$$

and assume the uniform bound $r \leq u_{n-1} \leq R$. Then we have the estimates:

$$|u_n - u_0| \leq \int_0^\tau |f_t(v_{n-1}, u_{n-1}) - u_{n-1}| \leq \tau D \leq D^{-1} q D = q.$$

So we have verified the uniform bounds $r \leq u_n \leq R$ for all n by induction. Since the estimates are independent of chioce of $x \in [x_0, x_0 + \tau]$, this verifies the claim.

Step 2: By step 1, both u_n and v_n have their ranges in $[r, R]$. So the Lipschitz constants $L_{\bar{f}}$ of $\bar{f}_t$ and $L_{\bar{g}}$ are both uniformly bounded at (u_n, v_n).

Let $V_n = (u_n, v_n)$. Then:

$$|V_{n+1} - V_n|$$
$$= |(\int_0^s (\bar{f}_t(v_n, u_n) - \bar{f}_t(v_{n-1}, u_{n-1})) dt, \int_{x_0}^x (\bar{g}_t(v_n, u_n) - \bar{g}_t(v_{n-1}, u_{n-1}) dy)|$$
$$\leq \max(s L_{\bar{f}} |V_n - V_{n-1}|, (x - x_0) L_{\bar{g}} |V_n - V_{n-1}|)$$
$$\leq \tau \max(L_{\bar{f}}, L_{\bar{g}}) |V_n - V_{n-1}| < \frac{1}{2} |V_n - V_{n-1}|.$$

Thus they are contracting:

$$\max\{|u_n - u_{n-1}|, |v_n - v_{n-1}|\} \leq \frac{1}{2^{n-1}} \max\{|u_1 - u_0|, |v_1 - v_0|\}$$

and $(u, v) = \lim_{n \to \infty} (u_n, v_n)$ exist uniformly. One also obtains the estimates:

$$\max\{|u_n - u|, |v_n - v|\} \leq \frac{1}{2^{n-1}} \max\{|u_1 - u_0|, |v_1 - v_0|\}.$$

Then they satisfy the integral equation:

$$(u, v) = (u_0, v_0) + (\int_0^s \bar{f}_t(v, u) dt, \int_{x_0}^x \bar{g}_t(v, u) dy)$$

which is equivalent to the hyperbolic Mealy system of the equations.

Step 3: Let us check that these satisfy the boundary conditions. The extension of u coincides with the one given along $\{x_0\} \times [x_0, x_0 + \tau]$, since the ordinary differential equation $u_s = \bar{f}_t(u, v_0)$ admits a unique solution with the given initial value $u(x_0, 0)$. u and v also satisfy the boundary conditions along $[x_0, x_0 + \tau] \times \{0\}$ and $\{x_0\} \times [0, \tau]$ respectively. So the pair (u, v) certainly satisfies the boundary condition. $\qquad\qquad\square$

Let us continue the construction of solution on $[0, \infty) \times [0, \infty)$.

(4) Let us extend solutions on D_m inductively by lemma 16.3. Then one obtains solutions on $[0, \infty) \times [0, \tau]$.

Since the initial values $u|[0, \infty) \times \{0\}$ and $v|\{0\} \times [0, \infty)$ take their ranges in $[r - q, R + q]$, the range of both u, v on $[0, \infty) \times [0, \tau]$ is also the same by lemma 16.2, since the pair (f_t, g_t) gives a stable Mealy dynamics over $[r, R]$.

Next let us regard that:

$$[0, \infty) \times \{\tau\} \cup \{0\} \times [\tau, 2\tau]$$

is a part of the boundary of $[0, \infty) \times [\tau, 2\tau]$, on which the boundary condition is given for the pair (u, v). By repeating the same process over $[0, \infty) \times [\tau, 2\tau]$, one obtains the solution $u, v : [0, \infty) \times [0, 2\tau]$ with the initial condition $u|[0, \infty) \times \{0\}$ and $v|\{0\} \times [0, 2\tau]$.

By iteration inductively, one finally obtains the solutions with the given boundary condition:

$$u, v : [0, \infty) \times [0, \infty) \to \mathbb{R}.$$

This completes the construction of solutions to the hyperbolic Mealy system.

Uniqueness:

Let us put the initial domain:

$$I_0 = [0, \infty) \times \{0\} \cup \{0\} \times [0, \infty).$$

Theorem 16.2. *Suppose that the pair (f_t, g_t) gives a stable Mealy dynamics over $[r, R]$, and give an initial condition:*

$$\begin{cases} u : [0, \infty) \times \{0\} \to [r + q, R - q] \\ v : \{0\} \times [0, \infty) \to [r + q, R - q]. \end{cases}$$

Then:

(1) there exists a positive solution:

$$u, v : [0, \infty) \times [0, \infty) \to (0, \infty)$$

which admit the uniform bounds:

$$r + q \le u(x, s), \ v(x, s) \le R - q.$$

(2) The solution is unique.

Proof. The first statement follows from the above construction.

Let us verify uniqueness. Suppose two solutions (u, v) and (u', v') exist with the same initial condition. Let:

$$a = |u - u'|, \ b = |v - v'| : [0, \infty)^2 \to [0, \infty).$$

Below we verify $a = b \equiv 0$. Notice that $a|[0, \infty) \times \{0\}$ and $b|\{0\} \times [0, \infty)$ both vanish. Firstly at $x = 0$, the ordinary differential equation:

$$u_s(0, s) = f_t(v(0, s), u(0, s)) - u(0, s)$$

admits the unique solution on $s \in [0, \infty)$, since $v(0, s)$ is the given initial value. So $u(x, 0) = u'(x, 0)$ holds for all $x \in [0, \infty)$. Then $v(0, s) = v'(0, s)$ holds by a parallel argument. In particular:

$$u(x, s) = u'(x, s), \quad v(x, s) = v'(x, s)$$

hold for all $(x, s) \in [0, \infty) \times \{0\} \cup \{0\} \times [0, \infty)$.

Both the solutions satisfy the integral equations:

$$\begin{cases} u(x, s) = u(x, 0) + \int_0^s \bar{f}_t(v, u)ds, \\ v(x, s) = v(0, s) + \int_0^x \bar{g}_t(v, u)dx. \end{cases}$$

Since $\bar{f}_t$ and $\bar{g}_t$ are Lipschitz, there exists L so that the estimates hold:

$$\begin{cases} a \le L \int_0^s (a + b), \\ b \le L \int_0^x (a + b). \end{cases}$$

Let us denote the union of the segments by:

$$\gamma_{x,s} = \{(\alpha, s) \cup (x, \beta) : 0 \le \alpha \le x, \ 0 \le \beta \le s\}.$$

Then the sum of the above inequalities gives the following inequality:

$$(a + b)(x, s) \le L \int_{\gamma_{x,s}} (a + b).$$

Let us choose (x, s) so that the estimate $L(x + s) < \delta < 1$ holds. Then let $D \subset [0, \infty)^2$ be the rectangle whose boundary is given by:

$$\partial D = \gamma_{x,s} \cup [0, x] \times \{0\} \cup \{0\} \times [0, s].$$

Let us choose some point $(x_0, s_0) \in D$ so that the equality $(a + b)(x_0, s_0) = \sup_{(p,q) \in D}(a + b)(p, q)$ holds. Then for $(x, s) \in D$, the estimates hold:

$$(a + b)(x, s) \leq (a + b)(x_0, s_0) \leq L \int_{\gamma_{x_0, s_0}} (a + b) \leq \delta \, (a + b)(x_0, s_0).$$

Since $\delta < 1$, this implies $(a + b)(x_0, s_0) = 0$ and hence $(a + b)(x, s) \equiv 0$ on D.

By changing the domains by parallel transport as in the above construction, we can follow the same argument as above. By iterating the same process, we conclude that $a = b \equiv 0$ holds on $[0, \infty)^2$. $\qquad\square$

16.7 Asymptotic C^1 estimates

Let us study C^1 estimates on solutions to hyperbolic Mealy systems, which is known as the energy estimates for hyperbolic equations. We shall present the estimates with the explicit constants.

For functions $u : [0, \infty)^2 \to [0, \infty)$, let us denote the semi-norm of the first derivatives by:

$$\|u\|_{\bar{C}^1} = \sup_{(x,s) \in [0,\infty)^2} \max \left\{ \, \left|\frac{\partial u}{\partial x}\right|(x, s), \, \left|\frac{\partial u}{\partial s}\right|(x, s) \, \right\}.$$

Remark 16.2. The reason why we use the notation $\bar{C}^1$ rather than just C^1 is that the former does not involve C^0 norm.

The following is elementary:

Lemma 16.4. *Let us give a discrete initial value:*

$$\begin{cases} u : \mathbb{N} \times \{0\} \to \mathbb{R}, \\ v : \{0\} \times \mathbb{N} \to \mathbb{R} \end{cases}$$

such that the estimates hold for all $n = 0, 1, 2, \ldots$:

$$|u(n + 1) - u(n)|, \ |v(n + 1) - v(n)| \ \leq \ \mu.$$

Then there is a constant C independent of the choice of the initial values and μ so that there is an extension of the initial value as:

$$\begin{cases} u : [0, \infty) \times \{0\} \to \mathbb{R}, \\ v : \{0\} \times [0, \infty) \to \mathbb{R} \end{cases}$$

equipped with uniform $\bar{C}^1$ bounds:

$$||u||\bar{C}^1([0, \infty) \times \{0\}), \quad ||v||\bar{C}^1(\{0\} \times [0, \infty)) \leq C\mu.$$

Proof. Let us extend the initial values by connecting the discrete values by segments. Then its C^1 approximations give the desired property.

□

Recall the number $D = D_{r,R}(f_t, g_t)$ in 16.6 and introduce another quantity:

$$B = \max(||(f_t)_u - 1||C^0, ||(f_t)_v||C^0, \ ||(g_t)_v - 1||C^0, ||(g_t)_u||C^0).$$

Let us give the exponential C^1 estimates:

Proposition 16.1. *Suppose that the pair (f_t, g_t) gives a stable Mealy dynamics over $[r, R]$.*

Let us give the initial condition:

$$u(\ ,0), \ v(0, \ ) : [0, \infty) \to [r + q, R - q]$$

with the uniformly bounded C^0 norms:

$$||u_x||C^0([0, \infty) \times \{0\}), \quad ||v_s||C^0(\{0\} \times [0, \infty)) \ \leq \ A < \infty.$$

Then there is a constant C so that the solution $u, v : [0, \infty) \times [0, \infty) \to (0, \infty)$ satisfies the asymptotic C^1 bounds:

(1) $\quad ||\dfrac{\partial u}{\partial x}||C^0([0, \infty) \times \{m\}), \ ||\dfrac{\partial v}{\partial s}||C^0(\{m\} \times [0, \infty))$

$\quad\quad \leq \ 2^{\tau^{-1}m}(A + 2D),$

(2) $\quad ||\dfrac{\partial u}{\partial s}||C^0, \ ||\dfrac{\partial v}{\partial x}||C^0 \ \leq \ D$

for all $m \in [0, \infty)$, where:

$$\max(\mathrm{Lip}_{\bar{f}_t}, \mathrm{Lip}_{\bar{g}_t}) \leq \frac{\tau^{-1}}{2}, \quad \tau \leq D^{-1}q, \quad \delta \equiv \tau B \leq \frac{1}{4}.$$

Proof. Let us use the notations in 16.6. Firstly we split the domains into periodic stripes. Then we verify some uniform estimates on the first derivatives of the approximated solutions (u_n, v_n) over each stripe, which are independent of n. Secondly we verify that they converge to the solutions uniformly in C^1 over each stripe.

Step 1: By theorem 16.2, the range of the solution (u, v) is uniformly bounded between r to R. Then it follows from the defining equations that both uniform bounds hold:

$$\|u_s\|C^0, \|v_x\|C^0 \leq D.$$

Moreover the solution is unique with respect to the given initial value.

Let us estimate u_x as follows. Recall the inductive construction of solutions over $[0, \infty) \times [0, \tau]$ in 16.6. We split the domain into periodic squares $D_m = [m\tau, (m+1)\tau] \times [0, \tau]$, and construct solutions successively on each D_m as $\lim_{n \to \infty}(u_n, v_n)$, where:

$$\begin{cases} u_n = u_0 + \int_0^s (f_t(v_{n-1}, u_{n-1}) - u_{n-1})dt, \\ v_n = v_0 + \int_{x_0}^x (g_t(v_{n-1}, u_{n-1}) - v_{n-1})dy \end{cases}$$

for $0 \leq s \leq \tau$ and $x_0 = m\tau \leq x \leq (m+1)\tau$.

Let us denote $u_n' = \frac{\partial u_n}{\partial x}$. Then:

$$u_n' = u_0' + \int_0^s ([(f_t)_u(v_{n-1}, u_{n-1}) - 1]u_{n-1}' + (f_t)_v(v_{n-1}, u_{n-1})v_{n-1}')dt,$$

$$v_n' = g_t(v_{n-1}, u_{n-1}) - v_{n-1}.$$

By lemma 16.3, the estimates hold:

$$|v - v_n|, \ |u - u_n| \leq \frac{1}{2^{n-1}} \max(|u_1 - u_0|, |v_1 - v_0|) \leq \frac{q}{2^{n-1}}.$$

Since the equality $v_x = g_t(v, u) - v$ holds, we obtain the estimates:

$$|v_n'| \leq |v'| + |v_n' - v'| \leq D + |(g_t(v, u) - v) - (g_t(v_{n-1}, u_{n-1}) - v_{n-1})|$$

$$\leq D + \frac{q}{2^{n-1}} Lip_{\bar{g}_t} \leq D + \frac{q}{2^n \tau}.$$

Let us put $\tau B = \delta \leq \frac{1}{4}$. Then we have the estimates:

$$|u_n'| \leq |u_0'| + \tau B \left[|u_{n-1}'| + D + \frac{q}{2^n \tau} \right]$$

$$= |u_0'| + \left(\delta D + B \frac{q}{2^n}\right) + \delta |u_{n-1}'| \leq \cdots$$

$$\leq (1 + \delta + \delta^2 + \ldots)(|u_0'| + D) + \left(\frac{1}{2^n} + \frac{1}{2^{n+1}} + \ldots\right) Bq$$

$$\leq 2\left(|u_0'| + D + \frac{Bq}{2^n}\right) \leq 2\left(A + D + \frac{Bq}{2^n}\right).$$

Similarly we have the estimates $|\frac{\partial v_n}{\partial s}| \le 2(A+D+\frac{Bq}{2^n})$ on $[0,\tau] \times [0,\infty)$. By construction, the estimates hold:

$$|\frac{\partial u_n}{\partial s}|, \quad |\frac{\partial v_n}{\partial x}| \le D.$$

Step 2: Let us verify C^1 uniform convergence of $\{u_n, v_n\}_n$. The Lipschitz constants of $(\bar f_t)_u$, $(\bar g_t)_v$ and $(f_t)_v$, $(g_t)_u$ are all finite and bounded on the range of u_n and v_n, by lemma 16.3. We assume that they are bounded by a constant B'.

Let us consider the estimates:

$$|v'_n| \le D, \quad |u'_n| \le 2\left(A + D + \frac{2Bq}{2^n}\right) \equiv \alpha_n.$$

on the domains $[0,\infty) \times [0,\tau]$. Then we have the estimates:

$$|(\bar f_t)_u(v_{n-1}, u_{n-1})u'_{n-1} - (\bar f_t)_u(v_{n-2}, u_{n-2})u'_{n-2}|$$
$$\le |(\bar f_t)_u(v_{n-1}, u_{n-1})u'_{n-1} - (\bar f_t)_u(v_{n-1}, u_{n-1})u'_{n-2}|$$
$$\quad + |(\bar f_t)_u(v_{n-1}, u_{n-1})u'_{n-2} - (\bar f_t)_u(v_{n-2}, u_{n-2})u'_{n-2}|$$
$$\le B|u'_{n-1} - u'_{n-2}| + B'(|u_{n-1} - u_{n-2}| + |v_{n-1} - v_{n-1}|)|u'_{n-2}|$$
$$\le B|u'_{n-1} - u'_{n-2}| + B'\alpha_{n-2}(|u_{n-1} - u_{n-2}| + |v_{n-1} - v_{n-2}|),$$

$$|(f_t)_v(v_{n-1}, u_{n-1})v'_{n-1} - (f_t)_v(v_{n-2}, u_{n-2})v'_{n-2}|$$
$$\le B|v'_{n-1} - v'_{n-2}| + B'(|u_{n-1} - u_{n-2}| + |v_{n-1} - v_{n-2}|)|v'_{n-2}|$$
$$\le B(g_t(v_{n-2}, u_{n-2}) - g_t(v_{n-3}, u_{n-3})| + |v_{n-2} - v_{n-3}|)$$
$$\quad + B'D(|u_{n-1} - u_{n-2}| + |v_{n-1} - v_{n-2}|)$$
$$\le B\tau^{-1}(|u_{n-2} - u_{n-3}| + |v_{n-2} - v_{n-3}|)$$
$$\quad + B'D(|u_{n-1} - u_{n-2}| + |v_{n-1} - v_{n-2}|).$$

By lemma 16.3, the estimates hold:

$$|u_n - u_{n-1}|, |v_n - v_{n-1}| \le \frac{q}{2^{n-1}}.$$

Thus we have the estimates:

$$|u'_n - u'_{n-1}| \le \tau[B|u'_{n-1} - u'_{n-2}|+$$
$$B'(\alpha_{n-2} + D)(|u_{n-1} - u_{n-2}| + |v_{n-1} - v_{n-2}|)]$$
$$+ B(|u_{n-2} - u_{n-3}| + |v_{n-2} - v_{n-3}|)$$
$$\le \delta|u'_{n-1} - u'_{n-2}| + \frac{C}{2^n}$$

for some constant C. Then we have the estimates:

$$|u'_n - u'_{n-1}| \leq \delta|u'_{n-1} - u'_{n-2}| + \frac{C}{2^n} \leq \frac{1}{4}|u'_{n-1} - u'_{n-2}| + \frac{C}{2^{n+1}}$$

$$\leq \frac{1}{2^4}|u'_{n-2} - u'_{n-3}| + \frac{C}{2^{n+2}} \cdots \leq \frac{1}{2^{n-1}}|u'_1 - u'_0| + \frac{C}{2^{2n-1}}.$$

Similarly we have the estimates:

$$|\frac{\partial v_n}{\partial s} - \frac{\partial v_{n-1}}{\partial s}| \leq \frac{1}{2^{n-1}}|v'_1 - v'_0| + \frac{C}{2^{2n-1}}$$

on $[0, \tau] \times [0, \infty)$.

So the convergence is uniform.

Step 3: By steps 1 and 2, we have the uniform estimates:

$$|\frac{\partial u}{\partial x}| \leq 2A + 2D$$

on $[0, \infty) \times [0, \tau]$, and:

$$|\frac{\partial v}{\partial s}| \leq 2A + 2D$$

on $[0, \tau] \times [0, \infty)$ by letting $n \to \infty$.

Let us repeat the same process of extensions of the solutions, on $[0, \infty) \times [\tau, 2\tau]$ for u and on $[\tau, 2\tau] \times [0, \infty)$ for v.

Notice that the initial norms have to replace A by $2A + 2D$. Then successively we have the estimates for $N = 0, 1, 2, \ldots$:

$$\begin{cases} |\frac{\partial u}{\partial x}| \leq 2^N A + (2^N - 1)2D & \text{on } [0, \infty) \times [(N-1)\tau, N\tau] \\ |\frac{\partial v}{\partial s}| \leq 2^N A + (2^N - 1)2D & \text{on } [(N-1)\tau, N\tau] \times [0, \infty). \end{cases}$$

So in total, we have the following estimates:

$$|\frac{\partial u}{\partial x}|(x, s) \leq 2^{\tau^{-1}s}(A + 2D)$$

$$|\frac{\partial v}{\partial s}|(x, s) \leq 2^{\tau^{-1}x}(A + 2D),$$

$$|\frac{\partial u}{\partial s}|, \quad |\frac{\partial v}{\partial x}| \leq D.$$

$\square$

Example 16.1. Let us consider the hyperbolic systems of the form:

$$\begin{cases} u_s = \frac{au}{1+u} - u, \\ v_x = g_t(u, v) - v. \end{cases}$$

Suppose an initial condition satisfies $u(1,0) = a - 1$. Then along the half line $\{(1,s) : s \geq 0\}$, the ODE $u_s = \frac{au}{1+u} - u$ has the unique solution $u(1,s) \equiv a - 1$. Then by differentiating the first equality by x variable, one obtains the equation $u_{xs} = (a^{-1} - 1)u_x$, whose solutions are given by:

$$u_x(1,s) = \exp((a^{-1} - 1)s)u_x(1,0).$$

They are uniformly bounded if $a > 1$ holds.

If we choose $a < 1$, u_x grows exponentially, even though u takes negative values.

Remark 16.3. If we we want to have better estimates in practical applications of C^1 estimates, then we can obtain exponential decay estimates, if we can induce some bounds such as $-aw(s) + b \leq w'(s) \leq -cw(s) + d$ for some positive $a, b, c, d > 0$. It can be given under some conditions of the defining rational functions f_t and g_t.

Let us induce the uniform energy estimates by assuming negative coefficients on derivatives as below.

Firstly we have the general estimates:

Lemma 16.5. *Let $w(s)$ satisfy the estimates:*

$$-aw(s) + b \leq w'(s) \leq -cw(s) + d$$

for some positive $a, b, c, d > 0$. Then the estimates hold:

$$\frac{b}{a} + \left(w(0) - \frac{b}{a}\right)\exp(-as) \leq w(s) \leq \frac{d}{c} + \left(w(0) - \frac{d}{c}\right)\exp(-as).$$

Proof. Let us rewrite the inequalities as:

$$-a\left(w(s) - \frac{b}{a}\right) \leq \left(w(s) - \frac{b}{a}\right)'.$$

Then we obtain the second inequality immediately.

The right hand side can be treated similarly. $\square$

Let us consider the hyperbolic system of PDE:

$$\begin{cases} u_s = f_t(v,u) - u, \\ v_x = g_t(v,u) - v. \end{cases}$$

Proposition 16.2. *Assume negativities:*

$$-a \leq (f_t)_u - 1, \quad (g_t)_v - 1 \leq -c$$

for some $0 < a, c$. Then the uniform estimates hold:

$$\frac{b}{a} + \left(u_x(x,0) - \frac{b}{a}\right)\exp(-as) \le u_x(x,s) \le \frac{b}{c} + \left(u_x(x,0) - \frac{b}{c}\right)\exp(-as),$$

$$\frac{d}{a} + \left(v_s(0,s) - \frac{d}{a}\right)\exp(-ax) \le v_s(x,s) \le \frac{d}{c} + \left(v_s(0,s) - \frac{d}{c}\right)\exp(-as)$$

where:

$$b = \sup|(f_t)_v(g_t - v)|, \quad d = \sup|(g_t)_u(f_t - u)|.$$

In particular $|u_x|$ and $|v_s|$ are both uniformly bounded.

Proof. Let us differentiate the defining equations:

$$\begin{cases} u_{xs} = ((f_t)_u - 1)u_x + (f_t)_v(g_t - v), \\ v_{xs} = ((g_t)_v - 1)v_s + (g_t)_u(f_t - u). \end{cases}$$

Then the conclusions follow by applying lemma 16.5. $\qquad\qquad\square$

16.8   Refinement and higher distortion

Let us consider a Mealy automaton:

$$\mathbf{A}: \quad \begin{cases} \psi : Q \times S \to S, \\ \phi : Q \times S \to Q \end{cases}$$

with two alphabets $S = \{s_0, s_1\}$, and embed them into the real number $S = \{L, L+1\} \subset \mathbb{R}$. Our aim is to verify the following:

Proposition 16.3. *Let $\mathbf{A}$ be a Mealy automaton with 2 alphabets.*
There is a refinement of $\mathbf{A}$ with the pair of functions $(\tilde{\phi}, \tilde{\psi})$ so that the corresponding elementary rational functions $(\tilde{f}_t, \tilde{g}_t)$ satisfy the estimates:

$$\{ |(\tilde{f}_t(v,u) - u)((\tilde{f}_t)_u(v,u) - 1)| + |(\tilde{f}_t)_v(u,v)||v_s| \}(x, s + \alpha)$$
$$< 2u(x, s+1),$$
$$\{ |(\tilde{g}_t(v,u) - v)((\tilde{g}_t)_v(v,u) - 1)| + |(\tilde{g}_t)_u(u,v)||u_x| \}(x + \alpha, s)$$
$$< 2v(x+1, s).$$

for any solution (u,v) and all $0 \le \alpha \le 1$.

The proof requires constructions of relative $(\max, +)$-functions by several steps and occupies the rest of sec. 16.8. We also induce some general estimates on rational functions with positive coefficients, which are used for our purpose later.

16.9 Prototype

Let us describe a prototype of rational functions, which we construct later by refinement.

Let $L \gg \delta > 0$ be positive numbers and $\xi : \mathbb{R} \to \mathbb{R}$ be a relative $(\max, +)$-function given by:

$$\xi(x) = \max(\min(x + \delta, L), x - \delta)$$
$$= \max(-\max(-(x + \delta), -L), x - \delta).$$

The orbits by ξ is increasing if the initial values satisfy $x < L$, and is decreasing if $x > L$. In particular ξ restricts to a self-map:

$$\xi : [L - q, L + q] \to [L - q, L + q]$$

for any $q > \delta > 0$.

Let f_t' be the corresponding elementary rational function to ξ:

$$f_t'(z) = t^{-\delta} z + \frac{t^L}{t^{L-\delta} + z} z.$$

Let f_t be tropically equivalent to f_t' given by:

$$f_t(z) = t^{-\delta} z + \frac{1}{N_0} \frac{t^L}{t^{L-\delta} + z} z$$

where $N_0 = N_0(t^\delta) \geq 1$ are chosen so that positivity holds:

$$\mu \equiv 1 - t^{-\delta} - \frac{t^\delta}{N_0} > 0.$$

It corresponds to the presentation:

$$\xi'(x) = \max(-\max(-(x + \delta), -L) - \max(0, \dots, 0), x - \delta).$$

Remark 16.4. We intend a case when $\delta > 0$ is small so that $t^\delta \sim 1$ is close to 1, while $t^L \gg 1$ is quite large.

Let us consider the equalities:

$$(f_t(z) - z)' = t^{-\delta} - 1 + \frac{1}{N_0} \frac{t^{2L-\delta}}{(t^{L-\delta} + z)^2},$$

$$f_t(z) - z = \left(t^{-\delta} - 1 + \frac{1}{N_0} \frac{t^L}{t^{L-\delta} + z}\right) z.$$

Then the estimates hold:

$$\frac{t^{2L-\delta}}{(t^{L-\delta} + z)^2}, \quad \frac{t^L}{t^{L-\delta} + z} \leq t^\delta.$$

Thus we have the estimates:

$$-1 < t^{-\delta} - 1 < (f_t(z) - z)' \le t^{-\delta} - 1 + \frac{t^\delta}{N_0} = -\mu < 0,$$

$$-z < (t^{-\delta} - 1)z < f_t(z) - z \le \left(t^{-\delta} - 1 + \frac{t^\delta}{N_0}\right) z = -\mu z.$$

So in total we have the estimates:

$$|(f_t(z) - z)'||f_t(z) - z| \le (1 - t^{-\delta})^2 \, z.$$

16.10   Construction of exit functions

Let us construct admissible pairs concretely. Such functions arise from *almost diagonal functions* as below. Let:

$$\xi(x) = \max(\min(x + \delta, L), x - \delta)$$

be the $(\max, +)$-function in 16.9. Given the initial value $x_0 = 0$, let us iterate it as $x_{n+1} = \xi(x_n)$. It is easy to see:

$$x_n \equiv L \qquad \text{if } x_0 = L, L+1 \text{ and } n \ge \delta^{-1}.$$

Let us consider another $(\max, +)$-function with 'two-step stairs' by:

$$\xi_2(x) = \max(\min(\xi(x), L+1), x - 3\delta).$$

This satisfies the properties:

$$\xi_2(x) \begin{cases} > x & x < L \\ = L & L - \delta \le x \le L + \delta \\ < x & x > L. \end{cases}$$

The orbits by iteration satisfy the same properties, $x_n \equiv L$ for all $n \ge \delta^{-1}$ for $x_0 = L, L+1$. On the other hand let us translate ξ_2. Then its properties change as:

$$\xi_2(x + 2\delta) \begin{cases} > x & x < L + 1 \\ = L + 1 & L + 1 - \delta \le x \le L + 1 + \delta \\ < x & x > L + 1. \end{cases}$$

In particular the orbits behave differently so that:

$$x_n \equiv L + 1 \qquad \text{if } x_0 = L, L+1 \text{ and } n \ge \delta^{-1}.$$

Inductively let us have the (max, +)-functions with n-step stairs by:

$$\xi_n(x) = \max(\min(\xi_{n-1}(x), L + n - 1), x - (2n - 1)\delta).$$

To represent exit functions, one can use ξ_l when we have l alphabets. In fact if we want an action which exchanges $L + a$ and $L + a + 1$ with $a < l$, then the translations:

$$\xi_l(\quad +2a\delta) : L + a, L + a + 1 \to L + a,$$
$$\xi_l(\quad +2(a + 1)\delta) : L + a, L + a + 1 \to L + a + 1$$

can play such roles. Notice that all ξ_n are 1-Lipschitz functions.

Let us fix n, and choose a large number $N_0 = N_0(n)$ which will be determined later.

Let f_t be the rational functions corresponding to ξ above:

$$f_t(z) = t^{-\delta}z + \frac{1}{N_0}\frac{t^L z}{z + t^{L-\delta}}.$$

For $0 \le m \le n$ let us put rational functions inductively by:

$$f_t^m(z) = t^{-(2m-1)\delta}z + \frac{1}{N_0}\frac{t^{L+m-1}f_t^{m-1}(z)}{t^{L+m-1} + f_t^{m-1}(z)}$$

which correspond to ξ_m defined above.

Lemma 16.6. *Let $N_0 \ge 1$ with the bound:*

$$t^{-\delta} + \frac{t^\delta}{N_0} \equiv 1 - \mu < 1.$$

Then the following estimates hold:

$$-1 + t^{-(2m-1)\delta} < (f_t^m(z) - z)' \le t^{-\delta} - 1 + \frac{t^\delta}{N_0} = -\mu < 0,$$

$$(-1 + t^{-(2m-1)\delta})z < f_t^m(z) - z \le (t^{-\delta} - 1 + \frac{t^\delta}{N_0})\, z = -\mu z.$$

In particular the estimates:

$$|(f_t(z) - z)'||f_t(z) - z| \le (-1 + t^{-(2m-1)\delta})^2\, z.$$

Proof. We have already verified the conclusion for $m = 1$.

Suppose f_t^{m-1} satisfies the conclusions for $m \le n$. Let us consider the equality:

$$(f_t^m)'(z) = t^{-(2m-1)\delta} + \big(\frac{t^{L+m-1}}{t^{L+m-1} + f_t^{m-1}(z)}\big)^2 \frac{(f_t^{m-1})'(z)}{N_0}.$$

Then we have the estimates:

$$(f_t^m)'(z) \le t^{-(2m-1)\delta} + \left(\frac{t^{L+m-1}}{t^{L+m-1} + f_t^{m-1}(z)}\right)^2 \frac{1-\mu}{N_0} < 1 - \mu,$$

$$(f_t^m)'(z) \ge t^{-(2m-1)\delta} + \left(\frac{t^{L+m-1}}{t^{L+m-1} + f_t^{m-1}(z)}\right)^2 \frac{t^{-(2m-3)\delta}}{N_0} > t^{-(2m-1)\delta}.$$

So we have the estimates $-1 + t^{-(2m-1)\delta} < (f_t^n(z) - z)' \le -\mu < 0$.

Next we have the estimates:

$$f_t^m(z) = t^{-(2m-1)\delta}z + \frac{1}{N_0}\frac{t^{L+m-1}}{t^{L+m-1} + f_t^{m-1}(z)}f_t^{m-1}(z)$$

$$\le t^{-(2m-1)\delta}z + \frac{1}{N_0}f_t^{m-1}(z) \le \left(t^{-(2m-1)\delta} + \frac{1-\mu}{N_0}\right)z$$

$$\le \left(t^{-\delta} + \frac{t^\delta}{N_0}\right)z < (1-\mu)z,$$

$$f_t^m(z) = t^{-(2m-1)\delta}z + \frac{1}{N_0}\frac{t^{L+m-1}}{t^{L+m-1} + f_t^{m-1}(z)}f_t^{m-1}(z)$$

$$\ge t^{-(2m-1)\delta}z + \frac{1}{N_0}\frac{t^{L+m-1}}{t^{L+m-1} + f_t^{m-1}(z)}t^{-(2m-3)\delta}z > t^{-(2m-1)\delta}z.$$

Thus we have verified the conclusions for m.

$\square$

Corollary 16.2. *Let N_0 be as above, Then solutions to the equation:*

$$u_s = f_t^n(u) - u$$

satisfy the following estimates:

$$|(f_t^n(u) - u)((f_t^n)_u(u) - 1)|(s+\alpha) < \frac{\tau^2}{1-\tau}u(s+1)$$

for $\tau = 1 - t^{-(2n-1)\delta}$.

Proof. Let $\tau = 1 - t^{-(2n-1)\delta}$. By lemma 16.6, the estimate $|f_t^n(z) - z| < \tau z$ holds.

Let us choose $0 \le \alpha_0 \le 1$ so that $\sup_{0 \le s \le 1} u(s) = u(\alpha_0)$ holds. Then we have the inequalities:

$$u(s+\alpha_0) \le u(s+1) + \int_{s+\alpha_0}^{s+1} |u_s| < u(s+1) + \tau \int_s^{s+1} u(a)da.$$

By the mean value theorem, we have the estimates:

$$\int_s^{s+1} u(a)da = u(s+\beta) \le u(s+\alpha_0).$$

Thus combining with these estimates, we obtain the following:

$$(1-\tau)u(s+\alpha_0) \le u(s+1).$$

Now finally we have the desired estimates:

$$|(f_t^n(u)-u)((f_t^n)_u(u)-1)|(s+\alpha) < \tau^2 u(s+\alpha)$$

$$\le \tau^2 u(s+\alpha_0) \le \frac{\tau^2}{1-\tau}u(s+1).$$

$\square$

Remark 16.5. $n\delta$ coincides with 1 up to some constant, and hence $t^{n\delta}$ is close to t. In order for the left hand side to be bounded by $2u(s+1)$, one has to choose $t > 1$ with $t^{(2n-1)\delta} < \sqrt{3}+2$.

In practical applications it will be desirable to have more effective constructions of rational functions which can be used for large $t \gg 1$.

16.11 Transition functions with 2 alphabets

Now let us consider a Mealy automaton:

$$\mathbf{A}: \quad \begin{cases} \psi : Q \times S \to S, \\ \phi : Q \times S \to Q. \end{cases}$$

Below we construct transition functions by relative $(\max, +)$-functions with two alphabets $S = \{s_0, s_1\}$. The reason for this restriction is rather technical, and just for simplicity of the notations. The general case can be considered similarly.

Let us embed S into the real number as $s_0 = L$ and $s_1 = L+1$ for some large $L \gg 1$. Let $Q = \{q^0, \dots, q^l\}$ be the set of the states. We prepare another copy of Q, and replace the set of states by the set of pairs of states:

$$Q = \{q^0, \bar{q}^0, \dots, q^l, \bar{q}^l\}.$$

Let us embed them as:

$$\bar{Q} \equiv \{q^0, \bar{q}^0, \dots, q^l, \bar{q}^l\} \subset \delta \, \mathbb{Z}$$

so that:

$$q^j = \bar{q}^j + 4\delta, \quad q^{j+1} = q^j + 8\delta$$

holds for $j = 0, \ldots, l$.

Lemma 16.7. *There exists a refinement of the pair (ψ, ϕ) by relative* $(\max, +)$-functions $\tilde{\psi}, \tilde{\phi} : \mathbb{R} \times \mathbb{R} \to \mathbb{R}$.

Proof. Let us construct each map separately.

Construction of $\tilde{\psi}$: We shall construct a $(\max, +)$-function which satisfies:

$$\tilde{\psi}(y, x) = \begin{cases} \xi_2(x) & |y - q^j| \le \delta, \\ \xi_2(x + 2\delta) & |y - \bar{q}^j| \le \delta. \end{cases}$$

Recall that for $x = L, L + 1$, $\xi_2(x)$ goes to L and $\xi_2(x + 2\delta)$ to $L + 1$ after iterating by several times.

Let us put the function:

$$\mu(y) = \min[2\delta, \max(\tau(y - q^0), \ldots, \tau(y - q^l))],$$
$$\tau(y) = \min[\max(0, y - \delta), \max(0, -y + 7\delta))].$$

It is straightforward to check the following properties:

$$\mu(y) = \begin{cases} 0 & |y - q^j| \le \delta, \\ 2\delta & |\bar{q}^j - y| \le \delta, \\ 0 & y \ge q^l + 7\delta \text{ or } y \le q^0 + \delta. \end{cases}$$

Now we put the desired functions by:

$$\tilde{\psi}(y, x) = \xi_2(x + \mu(y)).$$

Construction of $\tilde{\phi}$: $\tilde{\phi}$ concerns orbits by the states. Denote the involution by $\bar{*}^j = \bar{q}^j$ where $\bar{\bar{q}} = q$. For $* = q^j$ or $\bar{q}^j$, let ϵ_0 be the permutation between L and $L + 1$. Let:

$$\tilde{\phi} : \mathbb{R} \times \mathbb{R} \to [q^0, q^l + 7\delta]$$

be the relative $(\max, +)$-function which satisfies the following properties:

(1) $\tilde{\phi}(*, \delta l) = *$ if $\delta l \ne \mathbb{Z}$ for $l \in \mathbb{Z}$ and $* \in Q$.

(2) $|\tilde{\phi}(y, x) - y| \le k\delta$ for all $x \in \mathbb{R}$ and $y \in \mathbb{R}$ for some k.

(3) Suppose $\phi(q', s) = q$ and let ϵ_0 be the permutation between $s_0 = L$ and $s_1 = L + 1$. Then for $*' = q'$ or $\bar{q}'$:

$$\tilde{\phi}(*', L + 1) = \begin{cases} q & \text{if } \psi(q', \ ) = \epsilon_0, \\ \bar{q} & \text{if } \psi(q', \ ) = \text{id} \end{cases}$$

$$\tilde{\phi}(*', L) = \begin{cases} \bar{q} & \text{if } \psi(q', \ ) = \epsilon_0, \\ q & \text{if } \psi(q', \ ) = \text{id}. \end{cases}$$

If we choose δ^{-1} equal to some integer l_0, then these constructions give the refinement.

$\square$

Remark 16.6. $\tilde{\phi}$ can be close to the identity. We will not desire so 'wide' dynamics of $\tilde{\phi}$, since it concerns dynamics by the states, which is normally implicit in automata.

16.12 Estimates for elementary rational functions

Let us induce C^0 and C^1 estimates on orbits by elementary rational functions in terms of the corresponding $(\max, +)$-functions.

Let f_t correspond to φ, and fix $\bar{y}^0 \equiv (y_1, \ldots, y_{n-1}) \in \mathbb{R}^{n-1}$. Then we consider:

$$\varphi \equiv \varphi(\ , y_1, \ldots, y_{n-1}) : \mathbb{R} \to \mathbb{R}.$$

Assume that φ has the form:

$$\varphi(x) = L_0 + \alpha(x - x_0)$$

as a function in some neighborhood $I_0 = (x_0 - \tau, x_0 + \tau) \in \mathbb{R}$.

For $\bar{w}^0 = t^{\bar{y}^0} \equiv (t^{y^1}, \ldots, t^{y^{n-1}}) \in \mathbb{R}_{>0}^{n-1}$ and $z_0 = t^{x_0}$, let:

$$f_t \equiv f_t(\ , \bar{w}^0) : (0, \infty) \to (0, \infty).$$

Lemma 16.8. *Let M be the number of the components of f_t.*

(1) For any $t^{x_0 - \tau} \leq z \leq t^{x_0 + \tau}$, the estimate holds:

$$\left(\frac{f_t(z)}{t^{L_0} z_0^{-\alpha} z^\alpha}\right)^{\pm 1} \leq M.$$

(2) Suppose φ takes bounded values from both sides:

$$L \leq \varphi \leq L + a.$$

Then there exists some constant C so that the estimates:

$$\left|\frac{\partial f_t}{\partial z}\right|(z) \leq C t^a \frac{t^L}{z}$$

hold for all $t > 1$, where $C = C(\deg h_t, \deg k_t, M)$ with $f_t = t^L \frac{h_t}{k_t}$.

(3) Suppose $a > 0$. Then there exists an integer N so that the estimates:

$$t^{L-a} \leq N f_t(z) \leq t^{L+3a}$$

hold for all $t^{L-a} \leq z \leq t^{L+3a}$ and all sufficiently large $t \gg 1$.

Proof. (1) Let $x = \log_t z$ and $\varphi_t(x) = \log_t f_t(z)$. By lemma 3.4, the estimate holds for all $x \in I_0$:

$$|\varphi_t(x) - \varphi(x)| = |\log_t f_t(z) - \log_t(t^{L_0 + \alpha(x - x_0)})|$$
$$= \log_t\left(\frac{f_t(z)}{t^{L_0} z_0^{-\alpha} z^\alpha}\right)^{\pm 1} \leq \log_t M.$$

By removing $\log_t$ from both sides, one obtains the desired bound.

(2) Notice that f_t is a parametrized rational function. For each $t > 1$, f_t takes bounded values from both sides, since:

$$M^{-1} t^{|\varphi(x)|} \leq t^{|\varphi(x)|} t^{-|\varphi_t(x) - \varphi(x)|} \leq |f_t(z)| \leq t^{|\varphi(x)|} t^{|\varphi_t(x) - \varphi(x)|} \leq M t^{|\varphi(x)|}.$$

In particular the degree of f_t must be equal to 0 with respect to z. So the derivative of f_t with respect to z has negative degree.

Let us denote $f_t = t^L \frac{h_t}{k_t}$ by the quotient of elementary polynomials. Then by the above estimates, we obtain the uniform bounds:

$$M^{-1} \leq \frac{h_t}{k_t} \leq M t^a$$

where the coefficients of both h_t and k_t are rational in t. Notice that all the coefficients take positive values.

Suppose h_t has degree $n_0 \geq 1$ in z. Then its derivative satisfies the estimates:

$$0 \leq z \frac{\partial h_t}{\partial z}(z) \leq c(n_0) h_t$$

for some constant $c(n_0)$ which is independent of $t > 1$.

Now suppose $\deg h_t = \deg k_t = n_0$, and denote $h'_t = \frac{\partial h_t}{\partial z}$. Then the estimates hold:

$$\left|\frac{\partial f_t}{\partial z}\right|(z) = t^L\left|\left\{\frac{h'_t}{k_t} - \frac{h_t k'_t}{k_t^2}\right\}\right| \leq 2t^L M t^a \frac{c(n_0)}{z} = 2c(n_0) M t^a \frac{t^L}{z}.$$

(3) Firstly we show that there exists an integer N and some $L \leq L' \leq L + a$ so that the estimates:

$$t^{L'} \leq N f_t(t^L) \leq t^{L+a}$$

hold for all sufficiently large $t \gg 1$.

There is some L' so that the estimates $\left(\frac{f_t(t^L)}{t^{L'}}\right)^{\pm 1} \leq M$ hold by (1) for some $L \leq L' \leq L + a$ and all $t > 1$. Since $\frac{f_t(t^L)}{t^{L'}}$ is a rational function in $t > 1$, there is a limit:

$$M^{-1} \leq \lim_{t \to \infty} t^{-L'} f_t(t^L) \equiv \kappa \leq M.$$

One may assume the estimates $NM \leq t^a$ with $N = \lceil \kappa^{\pm 1} \rceil + 1$ for all large $t \gg 1$, where $\lceil \kappa^{\pm 1} \rceil = \max(\lceil \kappa \rceil, \lceil \kappa^{-1} \rceil) \geq 0$.

Then one obtains the estimates $1 \leq N t^{-L'} f_t(t^L) \leq NM \leq t^a$, and so:

$$t^L \leq t^{L'} \leq N f_t(t^L) \leq t^{L'+a} \leq t^{L+2a}$$

since $L \leq L' \leq L + a$ holds. It follows from (2) that the estimates:

$$NM^{-1}t^L \leq N f_t(t^x) \leq N f_t(t^L) + CN t^{L+a} \log t^{4a}$$

hold for all $L - a \leq x \leq L + 3a$.

One may assume the estimates for all sufficiently large $t \gg 1$:

$$N^{-1}M \leq t^a, \quad CN t^{L+a} \log t^{4a} \leq t^{L+2a}.$$

Then combining these estimates:

$$t^{L-a} \leq N f_t(t^x) \leq 2t^{L+2a} \leq t^{L+3a}$$

holds for all $L - a \leq x \leq L + 3a$.

$\square$

16.13 Proof of proposition 16.3

Proof. Step 1: Let $\tau(y)$ be as in 16.11 and l_t correspond to τ:

$$l_t(w) = [(1 + t^{-\delta}w)^{-1} + (1 + t^{7\delta}w^{-1})^{-1}]^{-1}.$$

The inequalities hold:

$$(1 + t^{6\delta})^{-1} \leq l_t(w)^{-1} \leq 2, \quad |l_t(w)|' \leq 4(1 + t^{-\delta}).$$

Let:

$$h_t(w) = K^{-1}[t^{-2\delta} + (l_t(t^{-q^0}w) + \cdots + l_t(t^{-q^l}w))^{-1}]^{-1}$$

which is tropically equivalent to the one corresponding to μ, where $K \in \mathbb{N}$ is chosen so that the estimate $h_t(w) \leq 1$ holds. Notice that the estimates hold:

$$[(l + 1)(1 + t^{6\delta})]^{-1} + t^{-2\delta} \leq (Kh_t(w))^{-1} \leq \frac{2}{l+1} + t^{-2\delta}.$$

By inceasing the number of l if necessary, one may assume the lower bound $1 - \chi \leq h_t(w)$ for sufficiently small $0 < \chi < 1$.

We have the estimates:

$$|h_t(w)'| = \frac{t^{-q^0 + 2\delta}}{K} \left| \frac{\Sigma_{j=0}^{l} t^{-(q^j - q^0)} l_t'(t^{-q^j} w)}{(t^{2\delta} + l_t(t^{-q^0} w) + \cdots + l_t(t^{-q^l} w))^2} \right|$$

$$\leq \frac{t^{2\delta}}{t^{q^0}} \frac{4}{K(l+1)} \sup_{w \in \mathbb{R}} |l_t'(w)| \leq CK^{-1} t^{-q^0}$$

where C is independent of $t > 1$.

Now let $\tilde{\psi}$ be as in 16.11, and recall ξ with corresponding f_t. Then corresponding to $\tilde{\psi}$ is the following:

$$\tilde{f}_t(w, z) = f_t(z h_t(w)).$$

By lemma 16.6, both the estimates hold:

$$(-1 + t^{-3\delta}) z < \tilde{f}_t(w, z) - h_t(w) z < -\mu z,$$
$$-1 + t^{-3\delta} < (\tilde{f}_t)_z(w, z) - h_t(w) < -\mu.$$

In particular we have the estimates:

$$-z < (-2 + t^{-3\delta} + h_t(w)) z < \tilde{f}_t(w, z) - z < -\mu z,$$
$$-1 < -2 + t^{-3\delta} + h_t(w) < (\tilde{f}_t)_z(w, z) - 1 < -\mu,$$
$$|(\tilde{f}_t)_w(w, z)| = |h_t'(w)| |z| |f_t'(z h_t(w))| \leq CK^{-1} t^{-q^0} z.$$

Step 2: Let $g_t(w, z)$ correspond to $\tilde{\phi}$. Let us denote $\bar{\phi}(y, x) = \tilde{\phi}(y, x) - y$. Then there is some a with the estimates $|\bar{\phi}(y, x)| \leq a\delta$ (see 16.11). So we have the equality $\bar{\phi}(y, x) = \max(\bar{\phi}(y, x), -a\delta)$.

Let $\bar{g}_t(w, z) = w^{-1} g_t(w, z)$ correspond to $\bar{\phi}$. Then $\bar{g}_t'(w, z) = \frac{\bar{g}_t}{N_0} + t^{-a\delta}$ is tropically equivalent to $\bar{g}_t$ for large $N_0 \in \mathbb{N}$. By lemma 16.6, $\bar{g}_t$ is uniformly bounded and so one may assume the estimates $-1 + \tau < \bar{g}_t'(w, z) - 1 < -\tau$ for some $0 < \tau < 1$. Then $g_t'(w, z) \equiv w \bar{g}_t'(w, z)$ is tropically equivalent to g_t and satisfies the estimates:

$$(-1 + \tau) w < g_t'(w, z) - w < -\tau w,$$
$$-1 + \tau < (g_t)_w'(w, z) - 1 < -\tau.$$

Moreover one may assume the estimate $|(g_t')_z| << 1$.

Step 3: We have to verify the estimates:

$$\{ |(\tilde{f}_t(v, u) - u)((\tilde{f}_t)_u(v, u) - 1)| + |(\tilde{f}_t)_v(u, v)| |v_s| \}(x, s + \alpha)$$
$$< 2u(x, s + 1),$$
$$\{ |(\tilde{g}_t(v, u) - v)((\tilde{g}_t)_v(v, u) - 1)| + |(\tilde{g}_t)_u(u, v)| |u_x| \}(x + \alpha, s)$$
$$< 2v(x + 1, s).$$

for any solution (u, v) and all $0 \le \alpha \le 1$.

To apply energy estimates in proposition 16.2, we need the bounds of the following quantities:

$$
\begin{cases}
-a \ \le (f_t)_u - 1, \ (g_t)_v - 1 \le \ -c \\
b = \sup |(f_t)_v (g_t - v)|, \\
d = \sup |(g_t)_u (f_t - u)|.
\end{cases}
$$

The first line has been verified by steps 1 and 2 with the constants $a = -1 + \tau$ and $c = \tau$ for some $0 < \tau < 1$, and $d << 1$ is small by step 2. If we choose a large $q_0 >> 1$, then one may assume that $b << 1$ is also small. In particular one may assume that both u_x and v_x have small norms by the energy estimates.

If both the bounds:

$$
\begin{cases}
|f_t(z, w) - z| < \tau z, \\
|(f_t)_u(z, w) - 1| < \tau
\end{cases}
$$

hold, then u satisfies the following estimates:

$$
|(f_t(u, v) - u)((f_t)_u(u, v) - 1)|(s + \alpha) < \frac{\tau^2}{1 - \tau} u(s + 1)
$$

by corollary 16.2. Similarly if the bounds $|g_t(z, w) - w| < \tau w$ and $|(g_t)_w(z, w) - 1| < \tau$ hold, then v satisfies the estimates:

$$
|(g_t(u, v) - v)((g_t)_v(u, v) - 1)|(s + \alpha) < \frac{\tau^2}{1 - \tau} v(s + 1)
$$

In particular if we choose $0 < \tau < 1$ so that $\frac{\tau^2}{1-\tau} < 2$ holds, then we have verified the desired estimates.

$\square$

Remark 16.7. Basically the construction of the refnement is quite general, and hence one can expect that for any pair of relative $(\max, +)$-functions (ϕ, ψ), there is a refinement $(\bar\phi, \bar\psi)$ by 1-Lipschitz functions so that the pair of the corresponding elementary rational functions $(\bar f_t, \bar g_t)$ is admissible.

What we have to do for our purpose of PDE analysis, is to find 'effective' pairs of admissible functions, which allows us to give nice globally analytic bounds of their solutions.

Bibliography

[AN] I. ADJAN AND S. NOVIKOV, *Infinite periodic groups I,II,III*, Izv. Akad. Nauk SSSR Ser. Mat. **32**, pp. 212–244, 251–524, 709–731 (1968).

[Ale] V. ALESHIN, *Finite automata and the Burnside problem for periodic groups*, Math. Zametki **11**, pp. 319–328 (1972).

[BG] L. BARTHOLDI AND R. GRIGORCHUK, *Hecke type operators and fractal groups*, preprint.

[BGH] R. BRYANT, P. GRIFFITHS AND L. HSU, *Hyperbolic differential systems and their conservation laws, Part I*, Selecta Math., New Series **1-1**, pp. 21–112 (1995).

[BKN] L. BARTHOLDI, V. KAIMANOVICH AND V. NEKRASHEVYCH, *On amenability of automata groups*, Duke. Math. J. **154-3**, pp. 575–598 (2010).

[Bri] J. BRIEUSSEL, *Dissertation*, Universite Paris 7, (2010).

[GKP] R. GRAHAM, D. KNUTH AND O. PATASHNIK, *Concrete mathematics*, Addison-Wesley (1994).

[GKZ] I. GELFAND, M. KAPRANOV AND A. ZELEVINSKY, *Discriminants, resultants and multidimensional determinants*, Birkhauser (1994).

[GNS] R. GRIGORCHUK, V. NEKRASHEVICH AND V. SUSHCHANSKII, *Automata, dynamical systems and groups*, ed. R. Grigorchuk, Dynamical systems, auomata, and infinite groups, Proc. Steklov Inst. Math. **231**, pp. 128–203 (2000).

[Gri] M. GRICK, *The pentagram map and Y-patterns*, Adv. Math. **227**, pp. 1019–1045 (2011).

[Grig1] R. GRIGORCHUK, *Burnside problem on periodic groups*, Funct. Anal. Appl. **14**, pp. 41–43 (1980).

[Grig2] R. GRIGORCHUK, *Degrees of growth of finitely generated groups and the theory of invariant means*, Izv. Acad. Nauk SSSR Ser. Mat. **48-5**, pp. 939–985 (1984).

[Gro1] M. GROMOV, *Hyperbolic groups*, in Essays in group theory, ed. S. Gersten, MSRI Publ. **8**, (1987).

[Gro2] M. GROMOV, *Asymptotic invariants of infinite groups*, London Math.Soc. LNS **182**, (1993).

[GZ] R. GRIGORCHUK, A. ZUK, *The lamplighter group as a group generated by* a *2-state automaton, and its spectrum*, Geom. Dedicata **87**, pp. 209–244 (2001).

[Har] P. DE LA HARPE, *Topics in geometric group theory*, Chicago Lectures in Mathematics (2000).

[Hir] R. HIROTA, *Nonlinear partial difference equations I*, J. of Phys. Soc. Japan **43**, pp. 1424–1433 (1977).

[HT] R. HIROTA AND S. TSUJIMOTO *Conserved quantities of a class of nonlinear difference-difference equations*, J. Phys. Soc. Japan **64**, pp. 3125–3127 (1995).

[Kat1] T. KATO, *Operator dynamics in molecular biology*, in the Proceedings of the first international conference on natural computation, Springer LNS in computer science **3611**, pp. 974–989 (2005).

[Kat2] T. KATO, *Interacting maps, symbolic dynamics and automorphisms in microscopic scale*, Int. J. Pure Appl. Math. **25-3**, pp. 311–374 (2005).

[Kat3] T. KATO, *Entropy comparisons and codings on interacting maps*, Kyoto University, preprint (2006).

[Kat4] T. KATO, *Deformations of real rational dynamics in tropical geometry*, GAFA **19**, No. 3, pp. 883–901 (2009).

[Kat5] T. KATO, *Dynamical scale transform and interaction graphs*, (in Japanese) RIMS proceedings **1650**, pp. 13–33 (2009).

[Kat6] T. KATO, *Geometric representations of interacting maps*, Int. J. Math. Math. Sci. Article ID 783738, 48 pages doi:10.1155/2010/783738 (2010).

[Kat7] T. KATO, *Pattern formation from projectively dynamical systems and iterations by families of maps*, in the Proceedings of the 1st MSJ-SI, Probabilistic Approach to Geometry, Advanced Studies in Pure Mathematics **57**, pp. 243–262 (2010).

[Kat8] T. KATO, *An asymptotic comparison of differentiable dynamics and tropical geometry*, Math. Phys. Anal. Geom. **14**, pp. 39–82 (2011).

[Kat9] T. KATO, *Automata in groups and dynamics and induced systems of PDE in tropical geometry*, J. Geom. Anal. **24-2**, pp. 901–987 (2014).

[Kat10] T. KATO, *A note on the pentagram map and tropical geometry*, arXiv 1405.0084 (2014).

[Kit] B. KITCHENS, *Symbolic dynamics*, Springer (1998).

[KP] T. KAPPELER AND J. PÖSCHEL, *KdV and KAM*, A Series of Modern Surveys in Math. **45**, Springer (2003).

[KT] T. KATO AND S. TSUJIMOTO, *A rough analytic relation on partial differential equations*, J. Math. Res. **4-4**, pp. 125–139 (2012).

[KTZ] T. KATO, S. TSUJIMOTO AND A. ZUK, *Spectral coincidence of transition operators, automata groups and BBS in tropical geometry*, to appear in Comm. Math. Phys.

[LM] G. LITVINOV AND V. MASLOV, *The correspondence principle for idempotent calculus and some computer applications*, Idempotency, ed. J. Gunawardena, Cambridge Univ. Press, pp. 420–443 (1998).

[Log] J. LOGAN, *An introduction to nonlinear partial differential equations*, John Wiley and Sons Inc. Publ. (2008).

[Mik] G. MIKHALKIN, *Amoebas and tropical geoemtry*, in Different faces of geometry eds, S. Donaldson, Y. Eliashberg and M. Gromov, Kluwer academic plenum publ., (2004).

[MS] W. DE MELO AND S. VAN STRIEN, *One dimensional dynamics*, Springer (1993).

[Nit] Z. NITECKI, *Differentiable dynamics*, The MIT Press (1971).

[OST1] V. OVSIENKO, R. SCHWARTZ AND S. TABACHNIKOV, *The pentagram map: a discrete integrable system*, Comm. Math. Phys. **299**, pp. 409–446 (2010).

[OST2] V. OVSIENKO, R. SCHWARTZ AND S. TABACHNIKOV, *Liouville-Arnold integrability of the pentagram map on closed polygons*, Duke Math. J. **162**, pp. 2149–2196 (2013).

[QRT] G. QUISPEL, J. ROBERT AND C. THOMPSON, *Integrable mappings and soliton equations II*, Physica D **34**, pp. 183–192 (1989).

[Sch1] R. SCHWARTZ, *The pentagram map*, Exp. Math. **1**, pp. 71–81 (1992).

[Sch2] R. SCHWARTZ, *The pentagram map is recurrent*, Exp. Math. **10**, pp. 519–528 (2001).

[Sch3] R. SCHWARTZ, *Discrete monodromy, pentagrams, and the method of condensation*, J. Fixed Point Theory Appl. **3**, pp. 379–409 (2008).

[Sin] Y. G. SINAI, *Probability Theory*, Springer-Verlag Berlin Heidelberg (1992).

[Sol] F. SOLOVIEV, *Integrability of the pentagram map*, Duke Math. J. **162**, pp. 2815–2853 (2013).

[TH] S. TSUJIMOTO AND R. HIROTA, *Ultradiscrete KdV equation*, J. Phys. Soc. Japan **67**, pp. 1809–1810 (1998).

[TM] D. TAKAHASHI AND J. MATSUKIDAIRA, *Box and ball system with a carrier and ultradiscrete modified KdV equation*, J. Phys. A: Math. Gen., **30**, pp. 733–739 (1997).

[TS] D. TAKAHASHI AND J. SATSUMA, *A soliton cellular automaton*, J. Phys. Soc. Japan, **59**, pp. 3514–3519 (1990).

[TTMS] T. TOKIHIRO, D. TAKAHASHI, J. MATSUKIDAIRA AND J. SATSUMA, *From soliton equations to integrable cellular automata through a limiting procedure*, Phys. Rev. Lett., **76**, pp. 3247–3250 (1996).

[Vir] O. VIRO, *Dequantization of real algebraic geometry on logarithmic paper*, Proc. of the European congress of Math. (2000).

[Wil] J. WILSON, *On exponential growth and uniformly exponential growth for groups*, Invent. Math. **155-2**, pp. 287–303 (2004).

[Woe] W. WOESS, *Random walks on infinite graphs and groups*, Cambridge tracts in mathematics **138**, (2000).

[Wol] S. WOLFRAM, *Cellular automata and complexity*, Addison Wesley (1994).

Index